NEEM

NEEM
The Divine Tree
Azadirachta indica

H.S. Puri
Herba Indica
India

CRC PRESS

Boca Raton   London   New York   Washington, D.C.

FIRST INDIAN REPRINT, 2012

This book contains information obtained from authentic and highly regarded sources. Reprinted material is quoted with permission, and sources are indicated. A wide variety of references are listed. Reasonable efforts have been made to publish reliable data and information, but the author and the publisher cannot assume responsibility for the validity of all materials or for the consequences of their use.

Neither this book nor any part may be reproduced or transmitted in any form or by any means, electronic or mechanical, including photocopying, microfilming, and recording, or by any information storage or retrieval system, without prior permission in writing from the publisher.

Direct all inquiries to CRC Press LLC, 2000 N.W. Corporate Blvd., Boca Raton, Florida 33431.

© 1999 CRC Press, LLC

Trademark Notice: Product or corporate names may be trademarks or registered trademarks, and are used only for identification and explanation, without intent to infringe.

Visit the CRC Press Web site at www.crcpress.com

Printed and bound in India by
Replika Press Pvt. Ltd.

ISBN 10 : 90-5702-348-2
ISBN 13 : 978-90-5702-348-4

FOR SALE IN SOUTH ASIA ONLY.

CONTENTS

	Foreword	vii
	Preface to the Series	ix
	Preface	xi
1	Introduction	1
2	Plant Sources	9
3	Chemical Constituents	23
4	Cultivation	33
5	Plant Raw Material	55
6	Quality Assurance of Plant Raw Materials	59
7	Wood for Fuel and Timber	61
8	Processing of Plant Raw Materials	65
9	Traditional Uses	77
10	Therapeutic Indications and Pharmacological Studies	87
11	In Veterinary Practice	111
12	Haircare and Bodycare Products	113
13	Toxicology	115
14	Neem in Agriculture	121
15	Neem Seed Cake as a Manure and Nitrification Inhibitor	129
16	Poultry and Cattle Feed	133
17	Neem and Pollution	141

18	Neem and Household Pests	145
19	Protection of Food Materials	151
20	Composite Plant for Utilization of Neem	159
21	Patents on Neem	163 |//
Index		167

FOREWORD

For medicinal and aromatic plants the clock has slowly, but surely, turned a full circle. The use of plants for health care started, as recorded in Indian and Chinese treatises available of that time, about 2 to 3 millennia before Christ and reached its zenith in the first millennium AD. These mention the use of plants and plant-based preparations in human and animal health care and occasionally for the preparation of household products. Thereafter for several centuries the plant materials occupied a pre-eminent position in the trousseau of a traditional practicing healer. It was only at the beginning of the second millennium that alchemie and mineral-based products also started appearing. The plants and herbals nonetheless continued to be widely used until the industrial revolution in Europe brought in synthetic products for various kinds of usages. These products, because of their ease of preparation and administration, led to a slide in the popularity of plant and herbal-based products. Among other reasons responsible for this was a belief (which still persists to a large extent) that most of the plant-based products/herbals are non-standard preparations and hence lack quality and efficacy. Noticeable batch-to-batch variations for the same products and lack of therapeutic consistency further eroded to some extent the credibility of these products.

At the present time, however, the increasing environmental degradation due to a burgeoning synthetic products industry has rung alarm bells the world over. Several scientists in various countries are now engaged in discovering or rediscovering the usefulness of plants and herbals for value-added products. Their quest for rediscovering the usefulness of plant materials has its basis in those leads or references which are mentioned in folklore or traditional systems as indigenous cures for several ailments. This resurgence of interest has enormous economic and commercial implications as well. However, at the same time the public at large and also scientists are conscious of the fact that if indiscriminately commercially exploited, this plant wealth may not last long. This has given rise to a paradoxical situation where, on the one hand, the public wants 'green' products, be it for medicinal use, personal hygiene or for its palate, but on the other, dwindling resources make us wary of environmental denudation. A balance has to be struck between demand and supply and in my view it can be best taken care of by sustained and structured 'social forestry' programs with an emphasis on planting those species which are proven sources of herbal drugs or phytomedicines.

If we dwell on this further, we find that this attitudinal change in the learned and lay public towards products originating from plant sources is basically because of a belief that 'green' products have distinct therapeutic advantages over allopathy in treating ailments like hepatitis, asthma, diabetes, arthritis, immune disorders, some tumours, etc. Likewise, cosmetics and biocides from plant sources are popular because they are soft for human use and ecofriendly. Commercial estimates indicate that 70 to 80 percent of the population in developing countries, accounting for over 50 percent of world population, depends partly or entirely on herbal remedies. According to *Indian Medicinal Plants: A Sectoral Study*, a report recently brought out by India's Exim Bank, the global trade in medicinal plants is estimated to be approximately

US$60 billion, of which India's share is about US$700 million. The world demand for herbal products has been quoted to be growing at a rate of 7 percent per annum. This growth in demand, especially in developing countries, is partly due to the ready availability of herbal remedies, a shortage of practitioners of modern medicines in many of these countries, and the socio-cultural background of the users. Farnsworth *et al.* (in *Bulletin of the World Health Organisation*, **63**, 965–85, 1985) have mentioned that even in developed countries, plant drugs are proving to be of great importance. In the USA, for example, 25 percent of all prescriptions contained plant extracts or active principles derived from higher plants. At this juncture, the importance of a plant like Neems come to the fore.

Neem, a large evergreen tree, commonly found throughout the Indo-Malaysia region, has been the subject matter of numerous scientific studies. Scientists the world over have carried out extensive work on its botanical, medicinal, industrial and agricultural usages. Practitioners of the Indian ayurvedic system advise the use of Panchang (five parts) of neem, i.e. leaves, bark, fruit, flower and root, for various applications. The seed is another extremely useful part, especially for its oil. Extracts of various parts of neem have proven medicinal properties—anthelmintic, antifungal, antidiabetic, antibacterial, antiviral, antifertility, etc. It is for these properties only that the practitioners of ayurvedic, *siddha*, *unani tibb* and homeopathic systems of medicine make extensive use of its parts. It is my firm belief that it is a only matter of time before even the allopathic system starts to make use of its medicinal properties in a regular way. Neem's use as an insecticide and pesticide is also well documented. There is no gainsaying the fact that its economic and commercial value lies in every single part having a proven utility.

I find this comprehensive treatise on Neem (*Azadirachta indica*) to be an excellent collation of the recorded observations, research efforts and accomplishments of scores of individual botanists, taxonomists, traditional medicine men, and medicinal chemists of past and present. It involves a massive effort on the part of the author wherein he has mapped the Neem in its entirety. Be it the botanical description, chemical constituents and products derived from its various parts, usage in human and animal healthcare, personal hygiene, household remedies or other commercially valuable products—the coverage has been extensive. As is observed in any monograph of this kind, it offers us a well-documented retrospect of the work carried out. The chapters are full of ideas which provide further leads to a researcher, good questions to an inquisitive mind, alternative ways to combat various ills to a physician, etc. I feel this monograph has come at the most appropriate time when everyone—scientists, the political system and citizens as part of non-governmental organisations—is concerned about dwindling natural resources and would like to have a direction on the sustainable use of what Mother Earth has provided.

Dr Naresh Kumar
Institute of Microbial Technology
Sector 39-A
Chandigarh, India

PREFACE TO THE SERIES

There is increasing interest in industry, academia and the health sciences in medicinal and aromatic plants. In passing from plant production to the eventual product used by the public, many sciences are involved. This series brings together information which is currently scattered through an ever increasing number of journals. Each volume gives an in-depth look at one plant genus, about which an area specialist has assembled information ranging from the production of the plant to market trends and quality control.

Many industries are involved such as forestry, agriculture, chemical, food, flavour, beverage, pharmaceutical, cosmetic and fragrance. The plant raw materials are roots, rhizomes, bulbs, leaves, stems, barks, wood, flowers, fruits and seeds. These yield gums, resins, essential (volatile) oils, fixed oils, waxes, juices, extracts and spices for medicinal and aromatic purposes. All these commodities are traded world-wide. A dealer's market report for an item may say "Drought in the country of origin has forced up prices".

Natural products do not mean safe products and account of this has to be taken by the above industries, which are subject to regulation. For example, a number of plants which are approved for use in medicine must not be used in cosmetic products.

The assessment of safe to use starts with the harvested plant material which has to comply with an official monograph. This may require absence of, or prescribed limits of, radioactive materials, heavy metals, aflatoxins, pesticide residue, as well as the required level of active principle. This analytical control is costly and tends to exclude small batches of plant material. Large scale contracted mechanised cultivation with designated seed or plantlets is now preferable.

Today, plant selection is not only for the yield of active principle, but for the plant's ability to overcome disease, climatic stress and the hazards caused by mankind. Such methods as *in vitro* fertilisation, meristem cultures and somatic embryogenesis are used. The transfer of sections of DNA is giving rise to controversy in the case of some end-uses of the plant material.

Some suppliers of plant raw material are now able to certify that they are supplying organically-farmed medicinal plants, herbs and spices. The Economic Union directive (CVO/EU No 2092/91) details the specifications for the **obligatory** quality controls to be carried out at all stages of production and processing of organic products.

Fascinating plant folklore and ethnopharmacology leads to medicinal potential. Examples are the muscle relaxants based on the arrow poison, curare, from species of *Chondrodendron*, and the antimalarials derived from species of *Cinchona* and *Artemisia*. The methods of detection of pharmacological activity have become increasingly reliable and specific, frequently involving enzymes in bioassays and avoiding the use of laboratory animals. By using bioassay linked fractionation of crude plant juices or extracts, compounds can be specifically targeted which, for example, inhibit blood platelet aggregation, or have antitumour, or antiviral, or any other required activity. With the assistance of robotic devices, all the members of a genus may be readily screened. However, the plant material must be **fully** authenticated by a specialist.

The medicinal traditions of ancient civilisations such as those of China and India have a large armamentaria of plants in their pharmacopoeias which are used throughout South East Asia. A similar situation exists in Africa and South America. Thus, a very high percentage of the World's population relies on medicinal and aromatic plants for their medicine. Western medicine is also responding. Already in Germany all medical practitioners have to pass an examination in phytotherapy before being allowed to practise. It is noticeable that throughout Europe and the USA, medical, pharmacy and health related schools are increasingly offering training in phytotherapy.

Multinational pharmaceutical companies have become less enamoured of the single compound magic bullet cure. The high costs of such ventures and the endless competition from me too compounds from rival companies often discourage the attempt. Independent phytomedicine companies have been very strong in Germany. However, by the end of 1995, eleven (almost all) had been acquired by the multinational pharmaceutical firms, acknowledging the lay public's growing demand for phytomedicines in the Western World.

The business of dietary supplements in the Western World has expanded from the Health Store to the pharmacy. Alternative medicine includes plant based products. Appropriate measures to ensure the quality, safety and efficacy of these either already exist or are being answered by greater legislative control by such bodies as the Food and Drug Administration of the USA and the recently created European Agency for the Evaluation of Medicinal Products, based in London.

In the USA, the Dietary Supplement and Health Education Act of 1994 recognised the class of phytotherapeutic agents derived from medicinal and aromatic plants. Furthermore, under public pressure, the US Congress set up an Office of Alternative Medicine and this office in 1994 assisted the filing of several Investigational New Drug (IND) applications, required for clinical trials of some Chinese herbal preparations. The significance of these applications was that each Chinese preparation involved several plants and yet was handled as a **single** IND. A demonstration of the contribution to efficacy, of **each** ingredient of **each** plant, was not required. This was a major step forward towards more sensible regulations in regard to phytomedicines.

My thanks are due to the staff of Harwood Academic Publishers who have made this series possible and especially to the volume editors and their chapter contributors for the authoritative information.

Roland Hardman

PREFACE

The study of herbs and alternative systems of medicine has experienced radical changes during the past decade. The developments in organic chemistry during the second world war and afterwards changed the lifestyle of most of the people of the world, and a feeling developed that every ill has a pill. This was not the end; projects for synthetic foods were on the way. These developments not only made the study of herbs and the other natural products outdated in the developed countries, but had a serious effect in the developing world as well, where centuries-old medicinal plant gardens were destroyed and agricultural practices of growing herbs ignored.

With the revival of interest in natural products and the 'back to nature' call, researchers started looking into the herbal literature of oriental civilisations, particularly those of India and China, in addition to European and other sources. It was gratifying to find that Harwood Academic Publishers also decided to publish books on the industrial profiles of well-known medicinal and aromatic plants, including those of the Ayurveda.

When Dr R. Hardman, the book series editor and my former guide at the University of Bath (UK), contacted me about writing a book on important Ayurvedic plants, I suggested Neem to him, to which he readily agreed. I was aware of the old medicinal virtues of this tree and the recently discovered insecticidal activity. Quite a number of books and other publications are available on the chemistry and agricultural uses of azadirachtin, the major constituent of neem. But the Western world knew little about the medicinal uses and the recent pharmacological and therapeutic researches into this important tree. I planned the book for these and other, not so well-known aspects. In view of the enormous amount of literature available, I decided to cover all the topics briefly, with a complete bibliography so that the interested reader may find the details from the original source.

I must thank Dr Naresh of the Institute of Microbial Technology (IMTECH), Chandigarh (India) and his staff for library facilities and technical help in the preparation of the manuscript. He also agreed to write the foreword to the book and critically corrected first proof. Ms Vibha provided information on the extraction process of azadirachtin. Dr Pushpinder S. Puri, Corporate Research Director, Air Products Inc. (USA) was helpful in linguistic revision of some parts of the book. The Research Foundation for Science Technology and Natural Resource Policy, India, is thanked for permission to print the poster regarding patent rights on Neem. The computer typing of the whole text, and the layout of the book was only possible with the assistance of my daughter Navneet and some help from my son Avneet. My wife Harminder made important contributions in some places.

H.S. Puri

1. INTRODUCTION

Azadirachta indica, called Neem or Nim in most parts of the world, is one of the very few trees known in the Indian subcontinent since antiquity. During excavation of the site at Mohandjodaro, now in Pakistan, which is as old as 2000 BC (Puri, 1969), in the era of proto-Australoid and proto-Dravidian culture, neem leaves were found. As is evident from Hindu mythology, on their arrival in the Indo-Gangetic plains, the Aryans also attached great importance to neem. They considered this tree to be of divine origin. It was said that *Amrita* (ambrosia, the elixir of immortality) was being carried by *Garuda* (demi-god of Hindu mythology which is part human, part bird) to heaven and a few drops of this *Amrita* fell on the neem tree. In another story, *Amrita* was sprinkled by *Indira* (the celestial king) on the earth, which gave rise to the neem tree, yet in another instance neem tree is related to *Dhanvantri* (the Aryan god of medicine). It was also mentioned that the sun god took refuge in the neem tree to escape from the awesome power of demons (Vijayalakshmi *et al.*, 1995).

The neem tree was considered a gift of God and *Sreva roga nivarni* (the panacea for all diseases). In an old proverb it was said,

> The land where the neem tree abound,
> Can death, disease there in be found ?

Neem has also been called "Heal all", "Divine Tree", "Village Pharmacy" and even "Nature's drugstore" (Rawat, 1995).

The ancient Indian found many therapeutic uses for the tree and also observed that the tree could survive in very dry and arid conditions. In due course of time, the name and fame of neem spread, not only in the remote areas of the Indian subcontinent but also in the adjoining countries in Asia, now known as Sri Lanka, Malaysia, Indonesia and Thailand. Since ancient times, India has had cultural and commercial relations with the people of these countries.

Whereas in folklore mainly the leaves and to some extent the oil was used in Ayurveda (the Indian system of medicine), Siddha (the system of medicine practiced in some parts of south India) and Unani Tibb (the Greco-Persian system of medicine), polyherbal preparations containing one, two or all five parts of the plant, i.e. leaves, bark, flower, fruit and root, called *panchang* in Ayurveda, were used. In the traditional systems of medicine, some of the preparations were for internal administration, while others such as nasal drops, medicated oils or fats were for external application.

European colonizers, on their arrival in India in the sixteenth century, also noticed this important tree and they called it *Margosa*. This term has been widely used in the subsequent literature and until recently neem was called *Margosa indica*, and neem oil was known as *margosa* oil. European physicians in India, as well as Indian physicians trained in the orthodox system of medicine (allopathy) and in homoeopathy, saw great virtues in the nineteenth century in the bark of neem both from the stem and the root, but mainly stem bark was used, because of its easy accessibility. The bark was considered a substitute for cinchona, widely prescribed for malaria and other fevers at that time. Neem bark was included in the Indian Pharmacopoeia, the Indian Homoeopathic Pharmacopoeia and even in the British Pharmaceutical Codex. At one time it was in the US National Formulary, but it is doubtful if the source of this drug was neem or the closely allied *Melia azedarach*, also called China berry, with which neem has very often been confused.

Keeping in view the importance of neem in Indian culture, some studies were carried out in the earlier part of the twentieth century to establish the therapeutic efficacy of the various claims made about it in the traditional systems of medicine. The researches showed that neem lacked profound pharmacological activity, which was considered important at that time for a herb to be a source of a drug. Neem was also not found effective against any disease, as compared to the other drugs available at that time. The oil with its foetid odor was not acceptable in any form, even for external application.

During the second world war, because of the scarcity of various raw materials and war needs, research work on the industrial utilization of neem oil started again (Siddiqui and Mitra, 1945a, 1945b, 1945c). These workers filed patents for the pharmaceutical use of neem bitters and for refining the oil. Mitra (1963) published a book on neem, under the auspices of the Oil Technological Institute in India, covering all aspects of information available at that time, which included history, post-harvest technology and processing of the seed to obtain a good quality oil, detailed chemistry and technology of the oil. The main aim of this book was to provide information to the general public on the proper use of neem products, particularly the use of oil for making soap and other industrial products.

Ketkar (1976) in the organisation Neem Mission, tried to popularize neem products, keeping in view the large number of trees growing in India, and the amount of oil and seed cake they can yield. The Neem Mission propagated the idea of making neem soap from the oil at the village level as a small-scale cottage industry, and the utilization of seed cake, left after the extraction of oil, as a manure and as a denitrifying agent for nitrogenous fertilizers. Due to this effort and that of other agencies, neem soap for toilet purposes became a household name in India.

It was well known to Indian farmers that locusts during invasion avoid the neem tree and that it has an antifeedant property. Pradhan and Jotwani (1962, 1968) brought this fact to the notice of the scientific world. Radwaski (1977, 1977a) gave a detailed account of the tree. Thakur *et al.* (1981) reviewed the literature. Slangen and Kerkhoff (1984) found the nitrification inhibitory activity in neem cake. Parmar (1987) gave an overview of neem research and use in India.

The research on neem got a new stimulus, when out of 2000 plants investigated for their action against insects, only neem gave promising results. It was found that it was not only effective against insects but also quite safe for human beings and other warm-blooded animals. The active compound was later isolated and identified as azadirachtin.

Azadirachtin attracted the attention of workers all over the world, and various studies were published on it; the most important of these are Warthen (1979, 1989), Marzu (1989), Kraus (1983), Ascher (1987, 1992), Morgan and Mandava (1987), Schmutterer (1987), Schmutterer and Ascher (1987), Siddiqui *et al.* (1988), Jacobson (1989), Ascher and Misner (1989) and Arnason and Philogene (1991). The commercialization of azadirachtin under the trade name *Margosan*-O and its clearance by the Environmental

Protection Agency (EPA) of the USA (Larson, 1987) started a new era of non-hazardous insect controlling agents from plants.

In due course of time, neem was found to be a multi-purpose tree (Ahmed and Grainge, 1986), which could be used for the day-to-day activity of human beings (Fig. 1). It was observed that neem could adapt itself to a dry, harsh and hostile climate and degraded soil, particularly in the dry arid regions of the world, where availability of water is quite poor. It could also be planted for soil reclamation (Sastry and Kanathekar, 1990). The tree could provide much-wanted shade to cattle and man in scorching heat and support undergrowth vegetation. The leaves could be used as fodder for ruminants, particularly at times of scarcity. During a recent drought in Gujarat, a west Indian state, a large number of cattle were saved by feeding them neem leaves. It acts as a wind-breaker, an avenue tree, and the dry leaves that fall on the ground provide organic matter for the soil to support vegetation. The wood can be used as fuel, so scarce in arid regions, and also as timber for household furniture, and for agricultural implements. The seed (Axtell and Fairman, 1992) can provide oil for use in household lamps for illumination, as a lubricant for agricultural machinery, against various pests and diseases and for soap. The oil when applied to leather goods prolongs their life and is also useful as a first-aid medicine for healing wounds and skin diseases of man and domestic animals. The seed cake, after washing, can be used in small amounts in poultry and cattle feed. It may be used as such as an organic manure. It not only provides nutrition to plants, but helps in the conservation of nitrogenous fertilizers and the elimination of nematodes.

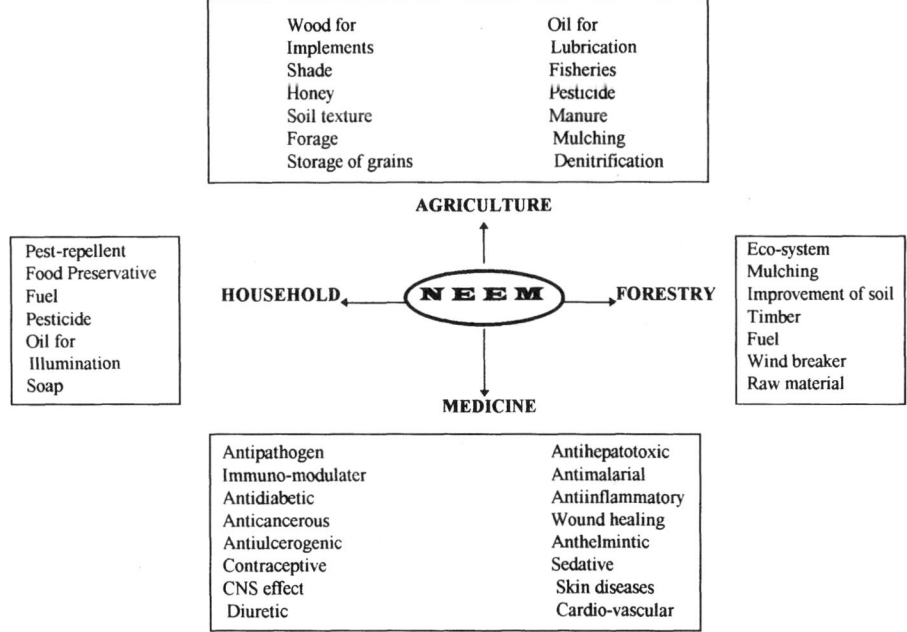

Figure 1 Possible uses of neem

International conferences on neem were held in Germany in 1980 and 1983 and in Kenya in 1986. In the proceedings of IUFRO (Salazar, 1990), cultivation of neem as a multi-purpose tree for arid zones was recommended. Koul *et al.* (1990) and Nat *et al.* (1991) in their review article dealt with various aspects of the chemical constituents of neem, followed by their chemistry and pharmacological activity. Hedin (1991) edited a book, based on two American Chemical Society symposia on naturally occurring pest regulators, which included neem. Isman *et al.* (1991) brought to light the various steps taken for the development of neem-based insecticide for Canada.

The book *Neem—a tree for solving global problems* was published by the National Research Council of the USA (Vietmeyer, 1992). Tewari (1992) in his book on neem, in addition to other details, gave an exhaustive account of botanical and forestry aspects, which were not available earlier. Under the auspices of the Indian Society of Tobacco Science, a symposium on the problems and prospects of botanical pesticides in integrated pest management was organized in 1990, in which the role of neem in agriculture was highlighted. In another seminar at Bangalore (India) in 1993, the theme was "Neem and Environment". Parmar and Singh (1993) also gave an account of the importance of neem in Indian agriculture, while Mordue and Blackwell (1993) presented an update on azadirachtin. Schmutterer and Doll (1993) described an allied species, *Azadirachta excelsa*, as a new source of insecticides. Read (1993) gave an account of the genetic improvement of neem at an international conference.

Randhawa and Parmar (1993) edited the book *Neem Research and Development* for the Society of Pesticide Science, India. Thomsen and Souvannavong (1994) described the activities of the international network on neem. Mariappan (1995) gave an account of neem in the management of crop diseases, while the journal *Indian Forester* came out with a special issue, "Neem—gift of the gods" (Khullar, 1995). Schumutterer (1995), who has done extensive research work on the pesticide activities of this tree, edited a book on neem. This book is a compilation of the research carried out by various workers all over the world, on all aspects of this wonderful plant. The book is a very authoritative source of information, and details have been given crop wise, as well as pest wise. An account of research carried out in applied entomology in the tropics is also available (Schmutterer, 1995a). Vijayalakshmi *et al.* (1995) published a user's manual for neem for household purposes and for small farmers.

The proceedings of the seminar held in Bangalore (India) in 1993 have been edited by Singh *et al.* (1995), while Kleeberg (1996) edited the proceedings of a workshop held in Germany. The book has chapters on the control of plant pests, including those found in greenhouses and others such as lice in children, fungicidal activity and the control of nematodes.

The University Grants Commission of India has awarded a book-writing project entitled "Neem—the wonder tree" to Dr M.D. Kharya. The book intends to cover those aspects not covered earlier (personal communication).

The publications mentioned above indicate the importance of this tree in the modern world. Exhaustive information is available now, particularly on the chemistry, pesticide activity and the role of neem in agriculture. Some topics such as history, detailed uses in Indian systems of medicine, botanical studies, mass propagation techniques, use for the preservation of food, pollution prevention, poultry and cattle feed, fertilizer, soil conservation, pharmacology, etc. have not been touched in most of the earlier publications; the details of these are given here. The well-known aspects like chemistry and use as a pesticide in agriculture have been touched on briefly.

REFERENCES

Ahmed, S.A. and Grainge, M. (1986) Potential of the neem tree for pest control and rural development. *Economic Botany*, **40**, 201–208.

Arnason, J.T. and Philogene, B.J.R. (1991) Symposium on the role of plant derived substances for insect control. *Memoires of Entomological Society of Canada*, **169**, 39–47.

Ascher, K.R.S. (1987) Notes on Indian and Persian lilac pesticide research in Israel. In H. Schmutterer and K.R.S. Ascher (eds.), Natural pesticide from neem tree (*Azadirachta indica A. Juss*) and other plants. *Proc. of 3rd Inter. Neem Conf.*, Nairobi, Kenya, 10–15 July, 1986, D.G.T.Z. Eschborn, Germany.

Ascher, K.R.S (1992) Anti-feedant, an overview. *Philippine Entomologist*, **8**, 117–123.

Ascher, K.R.S. and Misner, J. (1989) The effect of neem on insects affecting man and animal. In M. Jacobson (ed.) *Focus on Phytochemical Pesticides*. Vol. I *The Neem Tree*. CRC Press Inc., Florida, USA.

Axtell, B.L. and Fairman, R.M. (1992) Minor Oil Crops. Part I Edible oil, Part II Non-edible oil, Part III Essential oil. *FAO Agriculture Science Bulletin* No. **94**.

Hedin, P.A. (ed.) (1991) Natural occurring pest bioregulator. *Symposium* Sr. No. **449**, Washington.

Isman, M.B., Koul, O., Arnason, J.T., Stewart, J. and Sallocum, G.S. (1991) Developing a neem based insecticide for Canada. *Memoires Entomological Society, Canada*, No. **159**, 39–47.

Jacobson, M. (ed.) (1989) *Focus on Phytochemical Pesticides* Vol. I *The Neem Tree*. CRC Press Inc., Florida, USA.

Ketkar, C.M. (1976) *Utilisation of Neem (Azadirachta indica A. Juss) and its by products*. Report of the modified neem cake manurial project, KVIC, Bombay, India.

Khullar, Pankaj (1995) Special issue: Neem—gift of the gods. *Indian Forester*, **121**, 973–1086.

Kleeberg, H. (ed.) (1996) Practice oriented results on use and production of neem ingredients and phermones. *Proc. of the 4th Workshop*, Bordighera, Italy, 28th Nov.–1st Dec., 1994. Druck & Graphic, Giessen, Germany.

Koul, O., Isman, M.B. and Ketkar, C.M. (1990) Properties and uses of neem (*Azadirachta indica*). *Canadian Journal of Botany*, **68**, 1–11.

Kraus, W. (1983) Biological active compound from Meliacae. *Proc. 2nd Int. Conf. Chem., Biochem. Active Natural Products*, Budapest, Hungary.

Larson, R.O. (1987) Development of Margosan-O, a pesticide from neem seed. In H. Schmutterer, K.R.S. Ascher, (eds.) Natural pesticide from neem tree (*Azadirachta indica A. Juss*) and other plants. *Proc. 3rd Inter. Neem Conf.*, Nairobi, Kenya, 10–15 July, 1986.

Mariappan, V. (ed.) (1995) *Neem for Management of Crop Diseases*. Associated Publishing Co., New Delhi, India.

Marzu, U. (1989) Economic aspects of producing natural pesticides from neem and its use in agriculture. *Quarterly Journal of International Agriculture*, **28**, 48–56.

Mitra, C.R. (1963) *Neem*. The Oil Technological Institute, Hyderabad, India.

Morgan, E.D. and Mandava, N.B. (1987) *Handbook of Natural Pesticides*. Vol. III *Insect Growth Regulators*, Part B. CRS Press Inc., Florida, USA.

Mordue, A.J. and Blackwell, A. (1993) Azadirachtin: an update. *Journal of Insect Physiology*, **39**, 903–924.

Nat, J.M.van der, Sluis, W.G.van der, De Silva, K.T.D. and Labadie, R.P. (1991) Ethnopharmacognostical survey of *Azadirachta indica A. Juss* (Meliaceae). *Journal of Ethnopharmacology*, **35**, 1–24.

Parmar, B.S. (1987) An overview of neem research and use in India during the year 1983–86. *Proc. of 3rd Inter. Neem Conf.*, Nairobi, Kenya, 10–15 July, 1986.

Parmar, B.S. and Singh, R.P. (eds.) (1993) *Neem in Agriculture*. Indian Agriculture Research Institute, New Delhi, India.

Pradhan, S.A. and Jotwani, M.G. (1962) Neem seed deterrent to locusts. *Indian Farming*, 12, 7–11.

Pradhan, S.A. and Jotwani, M.G. (1968) Neem seed as insect deterrent. *Chemical Age, India*, 19, 756–760.

Puri, H.S. (1969) Classification of Drugs in Ancient India. *Everyday Science*, 24, 41–43.

Radwaski, S. (1977) Neem Tree. Part II Uses and potential uses. *World Crops and Livestock*, 29, 111–113.

Radwaski, S. (1977a) Neem Tree. Part III Further uses and potential uses. *World Crops and Livestock*, 29, 167–168.

Randhawa, N.S. and Parmar, B.S. (1993) *Neem Research and Development*. Publication No. 3. Society of Pesticide Sciences, India.

Rawat, G.S. (1995) Neem (*Azadirachta indica*) nature's drugstore. *Indian Forester*, 12, 977–980.

Read, M.D. (ed.) (1993) Genetic improvement of neem strategies for the future. *Proc. of an International Consultation held at Kasetart University, Bangkok, Thailand*, 18–22 January, 1993. Winrock International Institute for Agricultural Development. Forestry/Fuelwood Research and Development Project, Bangkok, Thailand.

Salazar, R. (ed.) (1990) Management and development of forest plantation using multipurpose species. *Proc. IUFRO meeting*, Guatemala, Centro Agronomico Tropical de Investigacion Ensenanza.

Sastry, T.C. and Kanathekar, K.V. (eds.) (1990) *Plants for Reclamation of Wasteland*. Publication and Information Directorate, CSIR, New Delhi, India.

Schmutterer, H. (1987) Model active substance for the environmentally safe control of injurious insects with special reference to substances contained in plants. (In German.) *Mitteilungen der Deutschen Gesellschaft fur Allagmeine und Angewandte Entomologie*, 5, 127–136.

Schmutterer, H. (1995) *Neem tree, source of unique natural products for integrated pest management, medicine, industry and other purposes*. VCH Verlag., Weinheim, Germany.

Schmutterer, H. (1995a) Results of research in applied entomology in the tropics, obtained by working group at Giessen University during the last ten years. (In German.) *Mitteilungen der Deutschen Gesellschaft fur Allgemeine und Angewandte Entomologie*. 10, 277–286.

Schmutterer, H. and Ascher, K.R.S. (eds.) (1987) Natural pesticides from neem tree (*Azadirachta indica A. Juss*)and other tropical plants. *Proc. 3rd Int. Neem Conf.*, Nairobi, Kenya, 10–15 July, 1986. D.G.T.Z. Eschborn, Germany.

Schmutterer, H. and Doll, M. (1993) The marrango or Philippine neem tree, *Azadirachta excelsa* (*A. integrifolia*) a new source of insecticide with growth regulating properties. *Phytoparasitica*, 21, 79–86.

Siddiqui, S. and Mitra, C. (1945a) Isolation of bitter principles from neem oil. *Indian Patent* No. 33649, 26 Nov 1945, 15 Nov 1946.

Siddiqui, S. and Mitra, C. (1945b) Separation of the bitter principles of neem oil and simultaneous refining of the oil. *Indian Patent* No. 33650, 26 Nov 1945, 15 Nov 1946.

Siddiqui, S. and Mitra, C. (1945c) Manufacture of new derivatives of subsidiary products, the physiologically active bitter principles of neem oil. *Indian patent* No. 33651. 26 Nov 1945, 16 Nov 1946.

Siddiqui, S., Siddique, B.S., Gaizi, S. and Mahmood, T. (1988) Tetracyclic triterpenoids and derivatives from Azadirachta indica. *Journal Natural Products*, 51, 30–43.

Singh, R.P., Chari, M.S., Raheja, A.K. and Kraus, W. (1995) *Neem and Environment*, Vol. I, II. International Book Distributors, Dehradun, India.

Slangen, J.H.G. and Kerkhoff, P. (1984) Nitrification inhibitor in agriculture and horticulture: a literature review. *Fertilizer Research*, 5, 1–76.

Tewari, D.N. (1992) *Monograph on Neem Azadirachta indica A. Juss*. International Book Distributors, Dehradun, India.

Thakur, R.S., Singh, S.B. and Goswami, A. (1981) *Azadirachta indica A. Juss* – A review. *Current Research in Medicinal and Aromatic Plants*, **3**, 135–140.

Thomsen, A. and Souvannavong, O. (1994) The International Neem Network. *Forest Genetic Resources*, **22**, 49–51.

Vietmeyer, N.D. (1992) *Neem: a Tree for Solving Global Problems*. National Research Council, Washington DC.

Vijaylakshmi, K., Radha, K.S. and Vandana Shiva (1995) *Neem – a user's manual*. Research Foundation for Science, Technology and Natural Resource Policy, New Delhi, India.

Warthen, J.D. Jr. (1979) *Azadirachta indica*: a source of insect and growth regulator. *US Dept. Agriculture Review Manual*. **ARM-NE4**, 21.

Warthen, J.D. Jr. (1989) *Azadirachta indica* organisms affected and reference list updated. *Proc. Entomological Soc. of Washington*, **91**, 367–388.

2. PLANT SOURCES

BOTANICAL NAMES

Azadirachta indica A. Juss
Syn. *Melia azadirachta* L Sp. Pl. I 385
   Melia indica Hooker, *Flora of British India*, I, 544.
   Melia indica Brandis, *Forest Flora*, 67

CLASSIFICATION

Family: Meliaceae, sub-family: Melioideae, Tribe: Melieae

Genus: *Azadirachta* A. Juss in Mem. Mus. Par. XIX (1830) 220 (Mel.68) 1830-Melia
Species: *indica* A. Juss Iicc 221, 69-Melia azadirachta

ORIGIN

According to Gamble (1902), the center of origin of *A. indica* is in the forests of Karnatka (south India) or the dried inland forests of Burma (Myanmar). Other authors were of the opinion that this tree originated in the forests of the Shivalik hills (foothills of the western Himalayas) or on east coast of south India. The great variety in the shape of the leaves and other morphological features support the theory of the origin of *A. indica* in upper Myanmar (Schmutterer, 1995); later it became naturalized in the forests of central and western India.

ETHNOBOTANICAL STUDIES

A detailed account of ethnobotanical studies is available in Watt (1889), Dymock *et al.* (1890), Dey (1896), and in many other old publications. In some of the recent botanical surveys in India, some more information has been collected by Billore and Audichya (1978), Tewari and Chaturvedi (1981) and Venkatraghavan and Sundersan (1981). Deka *et al.* (1983) reported it as an Ayurvedic plant from Assam, Sudhakar and Rao (1985) from the upper east Godavari district in Andhra Pradesh, Aminudin *et al.* (1993) from eight districts of Orissa and Singh (1994) documented its use by the tribal Kols of Uttar Pradesh.

In some surveys, stress was laid on finding the cure for certain diseases. Oommachan *et al.* (1990) reported the use of neem for jaundice and Aminudin *et al.* (1993) for malaria.

The use of neem in leather technology at the village level for preservation of leather goods is well known. Pal (1994) noted that it was also used for curing snake skin.

ETYMOLOGY

The present popular name "neem", also spelled earlier as "nim", has been derived from the Sanskrit word "nimba" which means sprinkler, which is the short term for "sprinkler of nectar (ambrosia)". The other Sanskrit synonyms for the tree, as given in the chapter on Ayurveda, refer to its habitat and the use of it in ancient India.

The meaning of the generic name *Azadirachta* does not appear to be interpreted properly in most of the literature. It is often said that it is from the Persian words *azad*—free, and *drakhat*—tree, i.e. free tree, and when the specific name *indica* is added to it, the meaning of the botanical name becomes the free tree from India, which does not convey any specific significance of the name.

The views expressed by Watt (1889) appear to be more convincing, according to which the Persians were well conversant with the allied tree *Melia azedarach*, also commonly known as the China berry, but in Persian as "*Azadarakhat*" (the corrupted form of it in most of the north Indian languages is *Dharek*). As discussed with a Persian scholar in Panjab University, India, *aza* means bitter in Persian and *drakhat* means tree, so the name of the China berry in Persian stood for "bitter tree". When the neem was introduced into Iran, to distinguish it from the China berry, which it resembled to a major extent, neem was called *Aza-drakhat Hindi*, i.e. the bitter tree from India, which led to the present botanical name *Azadirachta indica*.

NAMES IN OTHER LANGUAGES

In the Indian subcontinent, it is largely understood by the name neem or nimb, but in some areas, particularly in the south and east of India, there are regional names for the tree. These as well as the names from other parts of the world are given in Table 1.

RECENT BOTANICAL STUDIES

After the morphological and taxonomic description of neem, under *Melia azadirachta* by Hooker (1872) in *Flora of British India*, no serious efforts were made to investigate its taxonomy. Pennington and Styles (1975), and Pennington (1981) at the New York Botanical Garden, gave the generic monographs of the various members of Meliaceae.

Radwanski (1977) and Radwanski and Wickens (1981) described these aspects from an economic point of view. A monographic account in French is also available (Anonymous, 1988). Tewari (1992) and Vijayalakshmi *et al.* (1995) also described the morphology of the tree, and Sindhu Veerendra (1995) gave an account of variation within the species.

HABITAT

Neem is a large-sized evergreen tree (Fig. 2), but younger trees in dry localities may become leafless for a short period, and new leaves may appear in March–April,

Table 1 Name of neem in other languages

Name of the language	Geographical location	Name for neem
Arabic	Middle east	*Aza darakhul hind*
Assamese	India (east)	*Neem gach*
Bali	Indonesia	*Intran*
Bengali	India (east)	*Neem gach*
Burmese	Myanmar	*Thinbaw, Tamabin, Kamakha*
English	Europe	*Margosa, Indian Lilac Tree*
French	Europe	*Azadirae d'Inde, Margousier*
German	Europe	*Indischer Zedrach*
Kanad	India (south)	*Vepa, Bevy, Hebbevu, Kai bevu*
Konkan	India (south)	*Bena rooku*
Marathi	India (west)	*Limba, Nimbay, Neem*
Nigerian	Africa (west)	*Dogonyaro*
Oriya	India (east)	*Limbo, Nimbu, Kakopholo*
Persian	Iran	*Aza drakhat i hind*
Portuguese	Europe	*Amargosiera, Margosa*
Sinhali	Sri Lanka	*Kohunmba*
Tamil	India (south)	*Vembu, Veppam, Nimbamu*
Telugu	India (south)	*Vepa, Yeppa, Nimanuv*

which are pinkish green in color. The tree may grow up to a height of 20 m and a girth of 2.5 m.

(a) Stem

The color of the bark varies according to the part of the plant, its age and locality. The younger branches have a lighter color bark but in a mature trunk it may be grey to greyish black, rough, feebly fissured, and exfoliating. The inner surface of the bark is fibrous and pinkish brown. Small deposits of gum may be present on the stem in some places, but occasionally in some trees, which are quite old and in a humid climate, a fetid sap may be exuded from the trunk.

Bisht *et al.* (1993) described the branching pattern of a six-year-old tree.

(b) Root

It is normally dicotyledonous in nature, but in more than half of the population, vesicular–arbuscular mycorrhizal (VAM) infection is present due to *Glomus* and *Cigaspora* at 250 cm length. The intensity of infection varies with the availability of water. Neem appears to be a highly mycorrhizal-dependent species. Bala *et al.* (1989) concluded that a deep-rooted growth habit along with VAM infection may be a survival mechanism when competing for nutrients and water with shallow-rooted and fast-growing plant species. Mohan *et al.* (1995) carried out a survey of root and rhizosphere soil samples from nursery and plantation. It appears that VAM not only increases the nutrient uptake of the plant but also makes the tree tolerant to root diseases, transplant shock, toxicity of heavy metals and seasonal extremities like drought, etc. It has also been said

Figure 2 *Azadirachta indica*, an old neem tree in a religious place (*Nadha Sahib*) near Chandigarh, India. The author with his wife is in the extreme right corner. (Photo by Gurcharn Singh)

that *Phytophathora cinnamoni*, which is destroying neem worldwide, may be attacking those trees that lack mycorrhiza, as VAM forms a cover on the root and thus protects the plant from pathogens.

(c) Leaves

These are alternate, exstipulate, on a long slender petiole (Fig. 3), dorsal side darker green, ventral light in color, leaves 20–40 cm long, dense at the end of branches, alternate, leaflets 7–15, sometimes up to 17, variable in shape, particularly with respect to the central axis (Fig. 4). The leaves appear smooth but closer examination of young leaves near the shoot apex reveals the presence of resin-secreting glands (Figs. 5A and B). The lower portion of the leaf stalk is covered over with extra floral nectaries.

Leaves are bitter to the taste but Jacob (1941) observed a tree without these principles. Soni *et al.* (1981) observed the flattening of twigs and crowding of leaves with

Figure 3 *Azadirachta indica*, a small portion of branch

prominent ridges and furrows with a disturbed phyllotaxy. The leaflets are 2–7 cm long and 1–4 cm broad, imparipinnate, lanceolate, upper side bigger than the lower but it may vary within a population, often alternate, obliquely falcate, coarsely and bluntly serrate. The breadth of lamina and the degree of dentation on the margin of leaflets vary from locality to locality. In general, leaflets from dry arid areas have narrow lamina and sharp teeth along the sides.

A study of the herbarium at New York Botanical Gardens and Panjab University revealed that the trees from arid areas at the foothills of the Himalayas in general have narrow leaflets with sharp dentation in their serrate margin, as compared to the trees

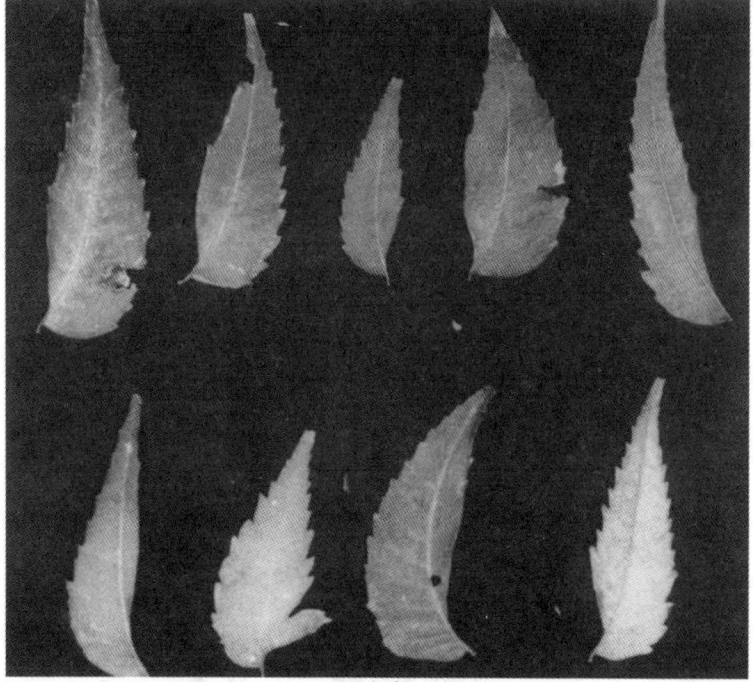

Figure 4 *Azadirachta indica*, leaflets from a single tree showing variations in the location of midrib, serration and other abnormalities

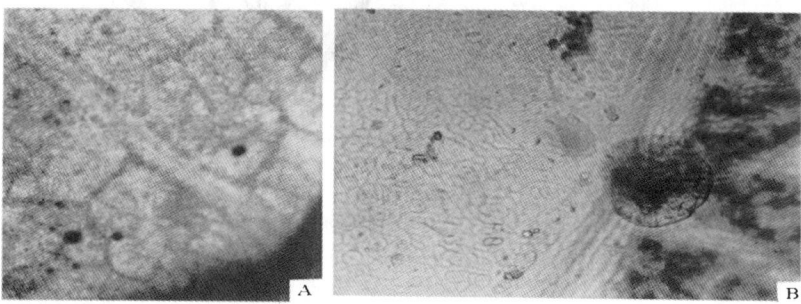

Figure 5 *Azadirachta indica*. A, resin glands on leaf surface, ×225; B, a gland enlarged ×800

from a humid climate which have comparatively broader leaflets with a less sharp serrate margin.

Sarma *et al.* (1992) studied the leaf architecture in relation to taxonomy. The major venation pattern was pinnate eucamptodemous. The authors have given a key for identification and differentiation of closely allied species on the basis of areoles, the presence and absence of bundle sheath and the leaf margin.

According to Schmutterer (1995), natural hybrids between *A. indica* and *A. siamensis* are found in upper Myanmar where both species grow together. The shape and consistency of leaflets in *A. indica* in this area is of intermediate type.

(d) Flowering

Trees growing in areas with a warm winter bloom first, followed by areas where the winter is comparatively cooler. Depending on the locality, flowering may range from January to May (Gupta *et al.*, 1995). Sporadic flowering in September–October has been observed quite often, in addition to that in February–March. Shanthi *et al.* (1996) reported abnormal seedlings in December from these trees and suggested that these trees may be used as germ plasm. In the Murshidabad area of West Bengal (India), the tree flowered throughout the year (Guhabakshi, 1984).

Generally the tree starts flowering at three to five years of age and becomes fully productive at the age of about ten years.

(e) The Flower

The buds are small, hermaphrodite, numerous, stalked, arranged in long, slender, very lax elongated axillary panicles, shorter than the leaves, bracts minute, deciduous. Flower buds open in the evening and are more scented at night. These buds give rise to 4–5 mm long whitish pink flowers.

Calyx: sepals 3 to 5 wide, imbricate, rounded, blunt, ciliate, sepals smooth and thin.

Corolla: petals 5, imbricate and oblong, oval in the bud, spreading, spathulate, somewhat twisted with a conduplicate claw, smooth outside, finely pubescent within. According to Gill *et al.* (1993) the number of petals may be 4–8, which is not genetically based.

Androecium: stamens 10, situated at the base of a hypogymous disk, the staminal filament combined into a long, cylindrical, erect tube, somewhat dilated below as well as at the top, furrowed and smooth externally, hairy within, terminating above in 10 blunt, thick, recurved trifid lobes, anther smooth erect, closely placed, introse, oblong, two celled. According to Pennington and Style (1975), the pollen grains are 3–4 corporate, prolate-spheroidal or sub-prolate, apocolpium medium, exine smooth, slightly thickened at the aperture.

Gynoecium: carpels, 3–5, syncarpous, superior, as many locule with 2 ovules in each loculus, style about the length of the staminal tube, stigma 5 lobed, placentation parietal (Nair, 1956). Garudamma (1956) and Rangaswamy and Promila (1972) have studied the embryology in detail. The ovary is trilocular at the base, becoming unilocular at the ovule-bearing region. The pollen tube is monosiphonous, and enters the ovule through a micropyle. One of the synergids is destroyed. Syngamy precedes triple fusion, resulting in an enlarged zygote. Twin embryos occur commonly (Nair and Kanta, 1961); the number of seed per embryo may be 1–3 (Gill *et al.*, 1993; Vijayan and Rehill, 1987). The gametophyte develops in the usual way; the embryo sac is of Polygonum type. Study of premature fallen fruits indicated that embryo abortion is common (Gill *et al.*, 1993).

Pollination

The flowers are cross-pollinated in general, in spite of bisexual flowers and the absence of self-incompatibility. Pollination is occasionally entomophilous but usually anemophilous.

Seed Development

A study of seed development has shown a steady increase in fruit/seed length, breadth, fresh and dry weight up to 12 weeks. The moisture content after this period started declining, with a increase in seed oil, protein and carbohydrates (Sivasamy and Karivaratharyu, 1993).

Fruit

It is an ovoid drupe, bluntly pointed, 1–2 cm long, when young and unripe smooth and green with white milky juice, yellow to brown when ripe, epicarp thin, mescocarp with scanty mucilaginous sweetish pulp, endocarp hard enclosing the seed. The fruit gets darker in color and wrinkled on maturity. Variability in seed size among different provenance was studied by Sindhu Veerendra (1995). The seed length varied between 11 and 18 mm, width 4.5–8.5 mm, and weight 100–530 mg. Among a three-year-old population, Gupta *et al.* (1995) observed that 92.3 percent of the trees had 1–100 fruits/tree, 4.36 percent 101–200 fruits/tree and 0.49 percent (one only) had more than 400 seed/tree.

Seed Dispersal

Most of the seed fall on the ground under the tree, where at that time the soil is water logged or there may be rain streams. The fruit may remain in moist conditions under the tree or occasionally may travel some distance with rain water. Since there is no dormancy, most of the seed may germinate immediately, but perish because of lack of conditions for further seedling growth. Occasionally, some fruits are swallowed by birds for their sweet pulp and the seeds are passed out of the body, undigested, because of the hard endocarp. The seeds so dropped are far away from the trees; if they germinate, the seedlings have much better chances of surviving and producing plants, as compared to undispersed seed. Gupta *et al.* (1995) studied the time of flowering and fruiting in a provenance trial of a neem population of three-year-old trees. In this population, 25.22 percent of trees exhibited fruits and 3.7 percent flowering, but in 0.25 percent, both flowering and the maturity of the fruits occurred at the same time.

Seedling Morphology

Germination is epigeal (Troup, 1921). Deb and Paria (1986) gave an account of seedling morphology. The fresh mature seed, if in humid conditions, may start germinating within a day or so, but fully dry seed may germinate from the 7th day of sowing and complete it in 25 days. Cotyledons are plano-convex, sub-opposite, lowest one sessile, blade obovate-oblong, stem erect. A few seedlings have an alternate cotyledon; in these, the lower cotyledon has a unilacunar two-trace node and the other a trilacunar three-trace node. In those cases where cotyledons are opposite, both cotyledons have a unilacunar two-trace or trilacunar three-trace condition (Bansal and Pillai, 1986). During the further course of development, the cotyledon along with the endocarp is pushed above the soil because of the elongation of the hypocotyl in the lower region. The plumule emerging from the cotyledon dislodges the endocarp. The top of the seedling at this stage is green, glabrous or with minute odorless glands. These glands are common in younger leaflets but become fewer in number as the leaf matures. The phenomenon of twin seedlings has also been observed, which may be as high as 11.27 percent (Vijayan and Rehill, 1987; Gurudev Singh *et al.*, 1995). This may be due to the development of one or more than one ovules, out of five of the ovaries (pentacarpellary), giving rise to more than one seed under the same endocarp, but according to Pushpakar and Babekey (1995) it is due to polyembryony, with a frequency of 1 out of 800.

Three mutants with white hypocotyles and leaves in a progeny of 1220 from a single tree have been observed by Kulkarni and Srimathi (1986, 1987). These seedlings survived for one month.

CYTOLOGY

The genetic chromosome number is n-14 and somatic 2n-28, 2n-30 have also been reported from the root tip mitosis (Pathak and Singh, 1949; Mukherji, 1952; Stylos and Vosa, 1971; Mehra *et al.*, 1972). According to Gill *et al.* (1993), this species has small-sized chromosomes but a high chromosome number. The loss in combination due to a low chiasmata index is compensated by an increase in the number of linkage groups and the allogamous nature. The open system of genetic recombination is also operating.

OTHER SPECIES

Two more species of *Azadirachta*, i.e. *A. excelsa* and *A. siamensis*, have been recognized (Schmutterer, 1995).

(a) *A. excelsa* (Jack) Jacobs (*A. integrifoliola* Merr.), sometimes erroneously spelled as *A. integrifolia*, is also called *marrango* or the Philippine neem tree. It was described as *Melia excelsa* Merill, and later as *A. integrifoliola*. *A. excelsa* is distinguished from *A. indica* by its entire leaflets, its panicle much longer than the leaves, and by its long flowers. Whereas *A. indica* thrives in hot dry regions, *A. excelsa* is a plant of lowland monsoon forests, and tolerates greater rainfall (Schmutterer and Doll, 1993). It is an endangered species because of over-exploitation in the Philippines.

(b) *A. siamensis* Val. is also called the Thai neem tree. It was identified earlier as *A. indica* var. *siamensis*, but Schmutterer (1995) recognized it as a distinct species, on the basis of a number of morphological, anatomical, histological, phytochemical and host infestation characters. In some areas, hybridization has occurred between *A. indica* and *A. siamensis*. The black neem reported from Thailand is probably a mutant of the Thai neem or a cross between the Thai and Indian neem (Schmutterer, 1995).

ANATOMICAL STUDIES

Both leaves and bark are medicinal, so have been subjected to detailed pharmacognostic studies, which mainly dealt with anatomical features. Wood as a source of timber has also been studied in detail by Brandis (1906), and recently by Tewari (1992). Some information about it has been provided in Chapter 7 on neem as a source of timber and fuel, in this book.

Naryana Aiyer *et al.* (1957), while dealing with the pharmacognostic studies of Ayurvedic medicinal plants, described both stem and root bark. Pandey (1969) presented the cuticular character of the leaf surface, while Farooqui (1981) gave an account of the epidermal structure. Purushothman *et al.* (1988) touched on both pharmacognostic and chemotaxonomic aspects of the bark, and compared it with the closely allied species *Melia azedarach*. Sarkar and Datta (1989) gave a morphological and anatomical account. Patel (1977) gave the diagnostic key for identification of bark powder, while Bagchi *et al.* (1992) studied the morphology of calcium oxalate crystal in neem and the other barks to distinguish them from each other.

Secretory Cells

Neem terpenes are present in all parts of living tissues but are most abundant in the seed kernel. The site of synthesis and accumulation of these has been identified as secretory cells. The study of secretory cells is thus of significance, because these accumulate triterpenoids. A new isoprenylated flavone was isolated from the resin gland and its structure was characterized by Balasubramanian (1993a).

Shah (1983), Inamdar *et al.* (1986), Arumugasamy *et al.* (1993) and Balasubramanian *et al.* (1993) gave an account of gum and resin secretions. Dayanandan *et al.* (1993), on the basis of microscopic and histochemical studies, established the following secretory cells:

1. Secretory cells that accumulate triterpenoids
2. Glandular trichomes, which secrete resin and protect young leaves
3. Gum ducts
4. Lacticifers which develop only in pericarp. These are articulate, and contain milky white latex.
5. Extrafloral nectaries

The secretory cells are found in the parenchymatous cells of the cortex and the pith of the stem, the petiole and cortex of the root, between the palisade and spongy tissue in leaves and all over the cotyledons. These cells are oval in shape and may contain pale-orange brown vesicles, which are age dependent as they become larger in the older cotyledons.

STUDIES ON GUM FORMATION

The neem tree yields a meager amount of gum, only in humid areas. Gum resin in many other trees is induced by a primitive method of acid treatment, which in many cases kills the plant by dehydration. Nair *et al.* (1980), Nair and Shah (1983) and Nair *et al.* (1985) tried ethephon and paraquat for this purpose. Ethephon administered to bark increased gum production by 100 percent. Histological and histochemical studies showed the disappearance of starch grains in the ethephon-treated plant, accompanied by the breakdown of cells and mass dissolution. The degradation products of all these tissues lead to gum formation. In heartwood treated with paraquat, gum ducts were observed, along with desiccation, the disappearance of starch grains and the accumulation of lipid insoluble polysaccharides and phenolics. When gum cavities were induced in sapwood by drilling holes, and treated with ethephon or paraquat, abundant high protein gum was produced after seven days in schizo-lysigenous cavities.

Nair (1988) studied wood anatomy and heartwood formation.

REFERENCES

Aminudin, Girach, R.D. and Khan, R.S. (1993) Treatment of malaria through herbal drugs from Orissa, India. *Fitoterapia*, **64**, 545–548.

Anonymous (1988) Azadirachta indica (*Azadirachta indica* A. de Jussieu). *Bois et Forets des Tropiques*, **217**, 33–47.

Arumugasamy, K., Udaiyan, K., Maniam, S., Balasubramaniam, C. and Mohan, P.S. (1993) Studies on the resin canal/ducts of Azadirachta indica. (abstr.). *Proc. World Neem Conference*, 24–28 February, 1993, Bangalore, India.

Bagchi, G.D., Srivastava, G.N. and Srivastava, A.K. (1992) Study on calcium oxalate crystals of some medicinal barks. *Indian Drugs*, **29**, 561–567

Bala, Kiran, Rao, A.V. and Tarafdar, J.C. (1989) Occurrence of VAM association in different plant species of the Indian desert. *Arid Soil Research and Rehabilitation*, **3**, 391–396.

Balasubramanian, C., Mohan, P.S., Arumugasamy, K. and Udaiyan, K. (1993) Chemical investigation on resin glands of Azadirachta indica A. Juss (abstr.) *Proc. World Neem Conference*, 24–28 February, 1993, Bangalore, India.

Balasubramanian, C., Mohan, P.S., Arumugasamy, K. and Udaiyan, K. (1993a) Flavonoid from resin glands of Azadirachta indica. *Phytochemistry*, **34**, 1194–1195.

Bansal, S. and Pillai, A. (1986) Variations in the cotyledonary node of Azadirachta indica A. Juss. *Acta Botanica Indica*, **14**, 54–56.

Billore, K.V. and Audichya, K.C. (1978) Some oral contraceptives—family planning in tribal ways. *Journal for Research in Indian Medicine, Yoga and Homeopathy*, **13**, 104–109.

Bisht, R.P., Toky, O.P. and Singh, S.P. (1993) Plasticity branching in some important tree species from arid north western India. *Journal of Arid Environment*, **25**, 307–313.

Brandis (1906) *Indian Trees* (Reprint, 1971) Bishen Singh Mohinderpal Singh, Dehradun, India, pp. 139.

Dayanandan, P., Stephen, A. and Murugandam, B. (1993) Location of neem terpenoids. (abstr.) *Proc. World Neem Conference*, 24–28 February, 1993, Bangalore, India.

Deb, D.K. and Paria, N. (1986) Seedling morphology of some economic trees. *Indian Agriculturist*, **30**, 133–142.

Deka, L., Majumdar, R. and Dutta, A.M. (1983) Some Ayurvedic important plants from district Kamrup, Assam, India. *Ancient Science of Life*, **3**, 108–115.

Dey, K.L. (1896) *Indigenous drugs of India*. Thacker Spink & Co, Calcutta, India.

Dymock, W., Warden, C.J.H. and Hooper, D. (1890) *Pharmacographica Indica* Vol. I. Kegan Paul, London.

Farooqui, P. (1981) Epidermal structure of some important forest plants, Neem Azadirachta indica A. Juss. *Indian Forester*, **107**, 237–242.

Gamble, J.S. (1902) *A Manual of Indian Timbers* (Reprint, 1972) Bishen Singh Mohninderpal Singh, Dehradun, India, pp. 143.

Garudamma, G.K. (1956) Studies in Meliaceae. I Development of embryo in Azadirachta indica A. Juss. *Journal Indian Botanical Society*, **35**, 222–225.

Gill, B.S., Singhal, V.K. and Maninder Kaur (1993) Studies on the genetic system of N-W Indian neem. (abstr.) *Proc. World Neem Conference* 24–28 February, 1993, Bangalore, India.

Guhabakshi, D.N. (1984) *Flora of Murshidabad District, West Bengal, India*. Scientific Publishers, Jodhpur.

Gupta, P.K., Pal, B.S. and Emmanuel, C.J.S.K. (1995) Initial Flowering and fruiting of neem in national provenance trials. *Indian Forester*, **121**, 1063–1068.

Gurudev Singh, B., Mahadevan, N.P., Shanthi, A.K.S. and Geetha, S.L. (1995) Multiple seeding in neem Azadirachta indica. *Indian Forester*, **121**, 1049–1052.

Hooker, J.D. (1872) *Flora of British India*, Vol. I. L. Reeve & Co., London.

Inamdar, J.A., Bhagvathi, Subramanian, R.B. and Mohan, J.S.S. (1986) Studies on the resin glands of Azadirachta indica A. Juss. (Meliaceae). *Annals of Botany, London*, **58**, 425–429.

Jacob, K.C. (1941) Margosa tree without the bitter principles. *Current Science*, **10**, 335.

Kulkarni, H.D. and Srimathi, R.A. (1986) Natural chlorophyll mutant in Azadirachta indica A. Juss. *Journal of Tropical Forestry*, **2**, 241–243.

Kulkarni, H.D. and Srimathi, R.A. (1987) An albino type natural chlorophyll mutant in Azadirachta indica A. Juss. *Silvae Genetica*, **36**, 44–45.

Mehra, P.N., Sareen, T.S. and Khosla, P. (1972) Cytological studies on Himalayan meliacae. *Journal of Arnold Arboretum*, **53**, 538–568.

Mohan, V., Verma, N. and Singh, Y.P. (1995) Distribution of VAM fungi in nurseries and plantation of neem (Azadirachta indica) in arid zone of Rajasthan. *Indian Forester*, **121**, 1069–1076.

Mukherji, S.K. (1952) Meiosis in Azadirachta indica A. Juss. *Current Science*, **21**, 287.

Nair, M.N.B. (1988) Wood Anatomy and heartwood formation in neem Azadirachta indica A. Juss. *Botanical Journal of Linnaeus Society*, **97**, 79–90.

Nair, M.N.B., Bhatt, J.R. and Shah, J.J. (1985) Induction of traumatic gum cavities in sap wood of neem (Azadirachta indica A. Juss.) by ethephon and paraquat. *Indian Journal of Experimental Biology*, **23**, 60–64.

Nair, M.N.B., Patel, K.R., Shah, J.J. and Pandalai, R.C. (1980) Effect of ethephon (2-chloroethyl phosphoric acid) on gumosis in the bark of Azadirachta indica. *Indian Journal of Experimental Biology*, **18**, 500–503.

Nair, M.N.B. and Shah, J.J. (1983) Histochemistry of paraquat treated wood in Azadirachta indica A. Juss. *IAWA Bulletin*, **4**, 249–254.

Nair, N.C. (1956) Placentation in Melia azadirachta L. *Current Science*, **25**, 264–265.

Nair, N.C. and Kanta, K. (1961) Studies on Meliacea IV. Morphology and embryology of Azadirachta indica A.Juss. A reinvestigation. *Journal Indian Botanical Society*, **40**, 382–386.

Naryana Aiyer, K., Nambooderi, A.N. and Kolammal, M. (1957) *Pharmacognosy of Ayurvedic Drugs*. Central Research Institute (Ayurveda), Trivandrum, India.

Oommachan, M., Shrivastava, J.L. and Masih, S.K. (1990) Observation of certain plants used in the treatment of jaundice. *Indian Journal of Applied and Pure Biology*, **5**, 99–102.

Pal, D.C. (1994) Tribal leather technology with special reference to reptile skin. *Ethnobiology in human welfare. Fourth Inter. Congress of Ethnobiology*, 17–21 November, 1994, Lucknow, India.

Pandey, Y.N. (1969) Studies on the cuticular characters of some meliaceae. *Bulletin Botanical Survey of India*, **11**, 377–380.

Patel, B.R. (1977) Diagnostic keys for Ayurvedic powdered crude drugs. *Indian Drugs*, **14**, 210–206.

Pathak, G.N. and Singh, B. (1949) Chromosome numbers in angiospermous plants. *Current Science*, **18B**, 347.

Pennington, T.D. (1981) A monograph of neo-tropical Meliaceae. In T.D. Pennington (ed.) *Meliaceae, Flora Neotropica*, Monograph No. 28. The New York Botanical Garden, New York, pp. 1–422.

Pennington, T.D. and Styles, B.T. (1975) A generic monograph of the Meliaceae. *Blumea*, **22**, 419–420.

Purushothaman, K.K., Sarda, A., Mathuram, V., Brindha, P., Sasikala E. and Rukmani, S. (1988) Pharmacognostic and chemotaxonomic studies of some Indian medicinal plants. *Acta Horticulturae*, **188A**, 165–167.

Pushpakar, B.P. and Babekey, G.S. (1995) A note on twin seedlings in neem Azadirachta indica A. Juss. *Indian Forester*, **121**, 1084.

Radwanski, S. (1977) Neem Tree 1. Commercial potential, characteristics, and distribution. 2. Uses and potential uses. 3. Further uses and potential uses. *World Crop and Livestock*, **29**, 62–66, 111–113, 167–168.

Radwanski, S. and Wickens, G.E. (1981) Vegetative fallows and potential value of neem tree, Azadirachta indica in the tropics. *Economic Botany*, **35**, 389–41.

Rangaswamy, N.S. and Promila (1972) Morphogenesis of the adult embryo of Azadirachta indica A. Juss. *Z. Pflanzen Physiology*, **67**, 377–379.

Sarkar, M.S. and Datta, P.C. (1989) Comparative pharmacognostic studies on the leaf drugs of Azadirachta indica and Melia azedarach. *Journal of Economic and Taxonomic Botany*, **13**, 236–240.

Sarma, V., Rao, S.R.S. and Beena, C.H. (1992) Leaf architecture in relation to taxonomy of Meliaceae. *Feddes Repertorum*, **7–8**, 535–542.

Schmutterer, H. (1995) *Neem Tree, source of unique natural products for integrated pest management, medicine, industry and other purposes*. VCH Verlag. Weinhim Germany.

Schmutterer, H. and Doll, M. (1993) The marrango or Philippine neem tree, Azadirachta excelsa (= A. integrifoliola): a new source of insecticides with growth regulating properties. *Phytoparasitica*, **21**, 79–86.

Shah, J.J. (1983) Gum resin and gum resin secretion in plants. *Acta Botanica Indica*, **11**, 91–96.

Shanthi, K., Manmuthu, L. and Gurdev Singh, B. (1996) Genetic significance of late flowering in neem. Azadirachta indica A. Juss as reflected by germination studies. *Indian Forester*, **122**, 263.

Sindhu Veerendra, H.C. (1995) Variation studies in provenances of Azadirachta indica (The neem tree). *Indian Forester*, **121**, 1053–1056.

Singh, J.P. (1994) Ethnobotany of Kols of Uttar Pradesh (abstr.). *Ethnobiology in human welfare. Fourth Int. Congress of Ethnobiology*, Lucknow, India, November, 17–21, 1994.

Sivasamy, M. and Karivaratharyu (1993) Tracing fruit and seed development and maturation by physical, physiological and biochemical methods in neem. Azadirachta indica A.Juss. (abstr.) *Proc. World Neem Conference*, 24–28 February, 1993, Bangalore, India.

Soni, K.K., Dadwal, V.S. and Jammaluddin (1981) A note on facitation in twigs of Azadirachta indica and Diospyros melanoxylon. *Indian Journal of Forestry*, **4**, 154–155.

Stylos, B.T. and Vosa, C.G. (1971) Chromosome number in the Meliaceae. *Taxon*, **20**, 485–499.

Sudhakar, S. and Rao, R.S. (1985) Medicinal plants of upper east Godavari district, Andhra Pradesh and need for establishment of medicinal farm. *Journal Economic and Taxonomic Botany*, **7**, 399–406.

Tewari, D.N. (1992) *Monograph on Neem Azadirachta indica A.Juss*. International Book Distributors, Dehradun, India.

Tewari, P.V. and Chaturvedi, C. (1981) Method of population control in ayurvedic classics. *Ancient Science of life*, **1**, 72–79.

Troup, R.S. (1921) *The Silviculture of Indian Tree*. Vol. I–III. Clarendon Press, Oxford, U.K.

Venkataraghvan, S. and Sundersan, T.P. (1981) Short note on contraceptive in Ayurveda. *Journal Scientific Research on Plants Medicinal*, **2**, 39–42.

Vijayan, A.K. and Rehill, P.S. (1987) Occurrence of twin seedlings in Azadirachta indica A. Juss. *Indian Journal of Forestry*, **10**, 214–216

Vijayalakshmi, K., Radha, K.S. and Shiva, V. (1995) *Neem: A User's Manual*. Research Foundation for Science, Technology and Natural Resource Policy, New Delhi, India.

Watt, J. (1889) *Dictionary of Economic Plants of India*. Superintendent, Government Printing Press, India.

3. CHEMICAL CONSTITUENTS

INTRODUCTION

The earlier Indian chemists concentrated on the bitter principles of neem oil. Sen and Banerjee (1931) gave an account of these. Siddiqui (1942) isolated the crystalline bitter compound nimbin, and later developed a method for separation of these bitters (Siddiqui, 1945a,b,c). Mitra (1963) described all these activities in his book on neem. According to Govindachari (1992), after the work on nimbin, interest in neem constituents became dormant except for the isolation of two new compounds, vilasinin and vepenin. With the discovery of the activity of the neem compound that suppressed feeding in locusts (Butterworth and Morgan, 1968), the detailed chemistry of these received a new stimulus. Very exhaustive studies were undertaken, particularly on the known active compound azadirachtin (Figure 6). Some of the important publications in this direction were by Lavie *et al.* (1971), Connolly (1983), Kraus (1983), Kraus *et al.* (1985, 1987), Morgan and Mandava (1987), Siddiqui *et al.* (1988) and Balandrin *et al.* (1988). Rastogi and Mehrotra (1990, 1991, 1995) in the *Compendium of Indian Medicinal Plants* in three volumes gave various stages of this study. In their monographic account of neem, Koul *et al.* (1990), Nat *et al.* (1991), Tewari (1992) and Govindachari (1992) gave a brief summary of the salient points of research on the chemistry of neem. Schmutterer (1995) has also given the relevant account of compounds which gave promising pesticidal results. Broadly speaking, the following types of compounds have been reported.

MAJOR CONSTITUENTS

Terpenoid constituents
(A) Protolimonoids
(B) Limonoids
(C) Pentatriterpenoids
(D) Hexatriterpenoids

Non terpenoid constituents
(A) Hydrocarbons
(B) Fatty acid
(C) Steroids
(D) Phenols
(E) Flavonoids
(F) Other compounds

Keeping in view the exhaustive literature on the chemistry of neem, for the present purposes only a brief summary of the research on chemistry has been given. The interested reader may get detailed information from the references given above.

SEED OIL

Dasa Rao and Seshadri (1942) and Child and Nathanael (1944) gave the fatty oil composition of oil. Skellon *et al.* (1962) and Mitra (1963) described the other compounds isolated from neem, most of which are now of historical importance. The important ones of these were nimbin mp 192° amorphous (1.2–1.6 percent) and nimbidin mp 90°. Hegnauer (1969) described sulphurous compounds. The most important compound isolated later on was azadirachtin.

AZADIRACHTINS

Butterworth and Morgan (1968) isolated azadirachtin (Figure 6), the active constituent, in crystalline form. The process of isolation included chromatographic fractionation of neem kernel extract, monitored by antifeedant assay with the desert locust, *Schistocera gregaria*. Butterworth *et al.* (1972) established the molecular formula for azadirachtin which was further modified by Kraus *et al.* (1985).

Remboldt *et al.* (1987b) determined the structure of azadirachtin B. The relationship between the structure and biological activity of azadirachtin A and B was established (Remboldt *et al.*, 1987a; Remboldt, 1989). Yamasaki and Klocke (1987) prepared eight derivatives of azadirachtin and bio-assayed them for their growth-inhibiting activity. The results showed that a hydroxyl group in azadirachtin was essential for the activity but for maximum action a lipophilic region is required. Broughton *et al.* (1986) also described the chemical structure. Barnby *et al.* (1989) and Remboldt (1989) studied the mode of action of the isomers of azadirachtin by exposing them to various wavelengths of ultraviolet radiation. These isomers were then analyzed for structural degradation and loss of biological activity. Further details on the pesticidal constituents can be had from Mordue and Blakewell (1993), Remboldt *et al.* (1993), Kobeleswaran *et al.* (1994) and Ley (1994).

Estimation of azadirachtin

Warthen *et al.* (1984) described a method of estimating azadirachtin in neem extract and formulations. Yamasaki *et al.* (1986) developed a rapid and inexpensive method for the

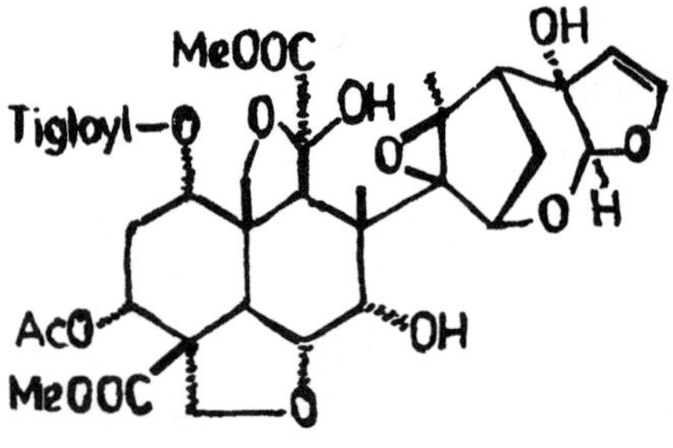

Figure 6 Structure of azadirachtin

isolation and purification of azadirachtin by using Flash Chromatography and High Performance Liquid Chromatography (HPLC). Schroeder and Nakanishi (1987) and Schneider and Ermel (1987) simplified the method of quantitative determination of azadirachtin. Ermel *et al.* (1987) determined this compound from various geographical regions of the world, and found a large variation in different countries. The highest concentration was from Nicaragua and Indonesia (4.7 percent) and the lowest from Sudan and Nigeria (1.5–1.9 per cent). Govindachari *et al.* (1990) developed a method of direct preparative HPLC without resorting to column chromatography. Govindachari *et al.* (1991, 1992, 1994) improved this method further to isolate azadirachtin A, B, D, H and I. The last two isomers were isolated for the first time and their structure was determined. Azadirachtin can also be estimated in other natural products (Sundaram and Curry, 1993) and in commercial preparation, where it is emulsified with surfactants, by a method of Azam *et al.* (1995).

Isolation of azadirachtins

On a small scale, azadirachtin may be isolated from the seed by the method given in Figure 7. The seed, after extraction with hexane, yields oil and a paste. This paste is first extracted with ethanol and then with chloroform to get crude azadirachtin, which may be about 20 gm from one kg of seed in some samples from India, but the yield varies from sample to sample.

Variation in azadirachtin in different samples

Twenty-one samples of neem seed, representing eight agro-ecological zones of India, were tested for azadirachtin and biological activity (Ketkar and Ketkar, 1993). In another study, azadirachtin varied from 0.05 to 4.24 percent in seed (Rengasamy *et al.*, 1993). Azadirachtin was also studied at different stages of flowering and fruiting by Rangaswamy and Parmar (1994); it was not detectable until forty days after anthesis but was present in green and yellow fruit. Yakkundi *et al.* (1995) detected azadirachtin in nine-week-old developing fruit; it started increasing with age and was at its maximum after the 19th week, when fruit started changing colour from green to yellow. Sidu and Behl (1996) studied seasonal variations in azadirachtins in seed in phenotypes that produced seed at the normal time in July–August (monsoon rains) and again in November–December. Monsoon seed yield 1.53 percent of azadirachtin-rich fraction as compared to winter seed (1.26 percent). Azadirachtin A was the major metabolite in the rainy season seed. Azadirachtin A and B were in equal proportion in winter seed. Concentration of azadirachtin F increased more than twofold in winter. Winter stress appeared to favor synthesis of azadirachtin B and F.

OTHER COMPOUNDS

Another important compound, salannin, has been studied by Henderson *et al.* (1968) and nimbin by Harris *et al.* (1968), who also described their stereochemistry. Gedunin, which has been found effective against the malarial parasite, was isolated by Lavie *et al.*

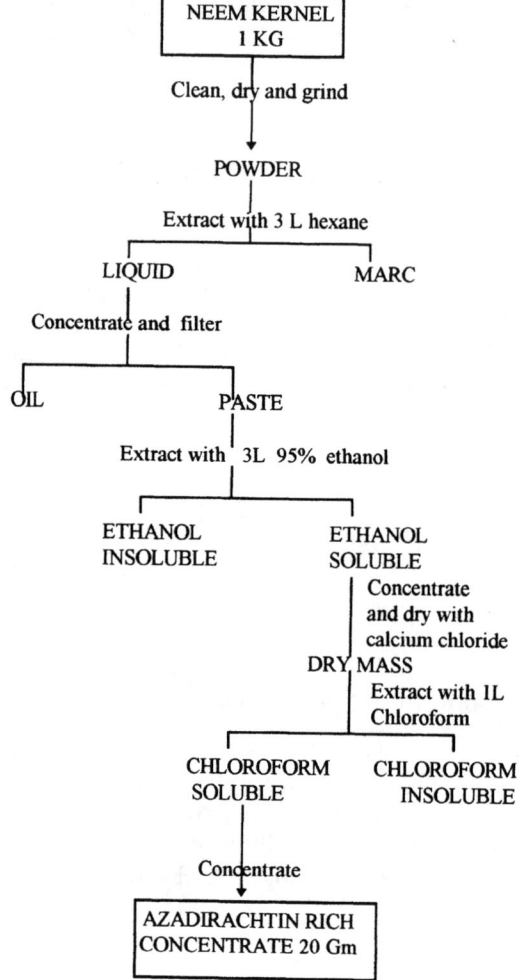

Figure 7 Flow diagram showing the method for extraction of azadirachtin

(1971). Yamasaki et al. (1988) described a method of isolation and purification of salannin which is 0.95 percent in seed oil.

CONSTITUENTS IN DIFFERENT ORGANS

Oil and seed cake

Skellon et al. (1962) described fatty acid of neem oil. Thakur and Godrej (1972) patented a method of purifying the oil. Keeping in view the importance of oil and oil cake, the Indian Standards Institute gave the specifications for both of these (Anonymous, 1968,

1977). The Solvent Extractor's Association of India (Anonymous, 1996) has also laid down standards for the oil. Sinha and Gulati (1968) and Singhal and Mudgal (1984) gave the amino acid content of neem seed cake.

Leaves

The smell of the leaves is due to essential oil (0.13 percent), as reported by Dakshinamurty (1954). The leaves have been investigated for other phytoconstituents also, the details of which can be had from Basak and Chakraborty (1968), Awasthi and Mitra (1971), Tirimanne (1984), Pant et al. (1988), Vashi and Patel (1988), and Katani and Padhya (1988).

Flower

The essential oil (0.025 percent) distilled from air-dried blossoms was reported by Subramanian and Rangaswamy (1947) to contain, besides tetrasulphides, kaempferol, thioamyl alcohol, benzyl alcohol, benzyl acetate and an unidentified alcohol.

Bark

Bitter principles, essential oils, and other constituents have been reported from bark, but the important compounds are polysaccharides, which are water soluble. These anti-inflammatory polysaccharides consist of glucose, arabinose and fructose in a molar ratio of 1:1:1 (Fujiwara, 1982, 1984). Subramanian and Lakshmanan (1993) reported the other constituents.

Heartwood

The wood is composed of cellulose, βcellulose, hemicellulose A, B and lignin (Wealth of India, 1948). Bhola Nath (1955) chemically examined the heartwood.

Gum

It is amber-coloured, non-bitter and gets blackened with age; it is water soluble and resembles gum acacia in some physical properties. The gum was investigated in detail by Pattabiraman et al. (1968). The gum is rich in protein. Bajpai et al. (1970) isolated pure aldobiuronic acid and aldotriuronic acids. The other studies were carried out by Anderson and Hendric (1971) and Anderson et al. (1972, 1986), who gave the amino acid and amino sugar composition.

Sap (stem exudate)

The sap has a strong smell of fermented liquor. It is slightly sweet and contains many amino acids (Wealth of India, 1948). Ali and Qadry (1988) investigated it in detail, and identified sugars (fructose, mannose and xylose), acids (citric, malonic, succinic, fumaric, and acetic), steroids (β sitosterol and methylenecycleartenol), limonoids (nimbin, azadirone, and gedunin), free amino acid (aminobutyric acid, glycine and a minor amount of argenine, glutamine, lysine and threonine) and crude proteins 3.57 g/100 ml.

REFERENCES

Ali, M. and Qadry, J.S. (1988) Studies on the stem exudate of Azadirachta indica. *Current Science*, **57**, 550–551.

Anderson, D.M.W., Bell, P.C., Gill, M.C.L., McDougal, F.J. and McNab, C.G.A. (1986) The gum exudate from Chloroxylon swietenia, Sclerocarya Caffra, Azadirachta indica and Moringa Oleifera. *Phytochemistry*, **25**, 249.

Anderson, D.M.W. and Hendric, A. (1971) The proteinaceous gum polysaccharide from Azadirachta indica A. Juss. *Carbohydrate Research*, **20**, 259–268.

Anderson, D.M.W., Hendric, A. and Munro, A.C. (1972) The amino acid and amino sugar composition of some plant gums. *Phytochemistry*, **11**, 733–736.

Anonymous (1968) *ISI Specifications of Neem Kernel Oil.* No. **4765**.

Anonymous (1977) *ISI Specifications for Neem cake for manuring.* No. **8558**.

Anonymous (1996) *Specifications of the oils.* The Solvent Extractor's Association of India, Bombay.

Awasthi, Y.C. and Mitra, C.R. (1971) Constituents of Melia indica leaves. *Phytochemistry*, **10**, 2842.

Azam, M.E., Seeni, R. and Parmar, B.S. (1995) Estimation of Azadirachtin content of emulsifiable and solution concentrate of neem. *Journal of AOAC International*, **78**, 893–896.

Bajpai, K.S., Chandrasekharan, V., Mukherjee, S. and Srivastava, A.N. (1970) Isolation of pure aldobiuronic acid and aldotriuronic acid from plant gum and mucilage. *Indian Journal of Chemistry*, **8**, 48–50.

Balandrin, M.F., Lee, S.M. and Klocke, J.A. (1988) Biologically active volatile organo-sulphur compounds from seed of neem tree, Azadirachta indica (Meliaceae). *Journal of Agriculture and Food Chemistry*, **36**, 1048–1054.

Barnby, A., Yamsaki, R.B. and Klocke, J.A. (1989) Biological activity of Azadirachtin (from Azadirachta indica), three derivatives and their ultra-violet degradation products on tobacco bud worm (Lepidoptera, Noctuidae) larvae. *Journal of Economic Entomology*, **82**, 58–63.

Basak, S.P. and Chakraborty, P.A. (1968) Chemical investigations of Azadirachta indica (M. azedarch) leaf. *Journal Indian Chemical Society*, **45**, 466–467.

Bhola Nath (1955) Chemical examination of the heartwood of Melia Azadirachta. *Journal of Scientific and Industrial Research*, **14B**, 634–636.

Broughton, H.B., Jones, P.S., Ley, S.V., Morgan, E.D., Salwin, A.M.Z. and Williams, D.J. (1987) Chemical Structure of Azadirachtin. In H. Schmutterer and K.R.S. Ascher (eds.) Natural Pesticides from neem tree (Azadirachta indica) and other tropical plants. *Proc. 3rd Inter. Neem Conf.*, Nairobi, Kenya, 10–15 July, 1987.

Butterworth, J.H. and Morgan, E.D. (1968) Isolation of a substance that suppresses feeding in locusts. *Chemical Communication*, **23–24**.

Butterworth, J.H., Morgan, E.D. and Perecy, G.R. (1972) Structure of functional groups. *Chemical Society, Perkin Transactions*, **1**, 2245–2250.

Child, R. and Nathanael, W.R.N. (1944) On the fatty acids of margosa (neem) oil. *Journal Indian Chemical Society*, **21**, 35–37.

Connolly, J.D. (1983) Chemistry of the limonoids of the meliaceae and ceneoraceae. In P.G. Waterman, and M.F. Grundon (eds.) Chemistry and chemical taxonomy of rutales. *Annual Review of Phytochemical Society of the Europe*, No. **22**, 175–213, Academic Press, New York.

Dakshinamurty, K. (1954) The amino acids in the leaf of Azadirachta indica (Melia). *Current Science*, **23**, 125–126.

Dasa Rao, C.J. and Seshadri, T.R. (1942) Fatty acids of neem oil. *Proceedings Indian Academy of Science*, **15**, 161.

Ermel, K., Pahlich, E. and Schmutterer, H. (1987) Azadirachtin content of neem kernel from different geographical location and its dependence on temperature, relative humidity and light.

In H. Schmutterer and K.R.S. Ascher (eds.) Natural pesticides from neem tree (Azadirachta indica A. Juss.) and other tropical plants. *Proc. 3rd Inter. Neem Conf.*, Nairobi, Kenya, 10–15 July, 1987.

Fujiwara, T., Sugishta, E., Takeda, T., Orihara, Y., Shimizu, M., Nomura, T. and Tomita, Y. (1984) Further studies on the structure of polysaccharides from the bark of Melia azadirachta. *Chemical Pharmaceutical Bulletin*, **32**, 1385–1391.

Fujiwara, T., Takeda, T., Ogihara, Y., Shimizu, M., Nomura, T. and Tomita, Y. (1982) Studies on the structure of polysaccharides from the bark of Melia azadirachta. *Chemical Pharmaceutical Bulletin*, **30**, 4025–4030.

Govindachari, T.R. (1992) Chemical and biological investigations on Azadirachta indica (the neem tree). *Current Science*, **63**, 117–122.

Govindachari, T.R., Gopalkrishna, G., Raghunathan, R. and Rajan, S.S. (1994) Crystallisation of azadirachtin A. *Current Science*, **66**, 295–297.

Govindachari, T.R., Sandhya, G. and Ganeshraj, S.P. (1990) Simple method for isolation of azadirachtin by preparative high performance liquid chromatography. *Journal of Chromatography*, **313**, 389–391.

Govindachari, T.R., Sandhya, G. and Ganeshraj, S.P. (1991) Isolation of novel azadirachtin H and I by high performance liquid chromatography. *Chromatographia*, **341**, 303–305.

Govindachari, T.R., Sandhya, G. and Ganeshraj, S.P. (1992) Azadirachtins H and I two neem tetranorterpenoids from Azadirachta indica. *Journal of Natural Products*, **55**, 596–601.

Harris, M., Henderson, R., McCrindle, R., Overton, K.H. and Turner, D.W. (1968) Tetranortriterpenoids-VIII. The constituents and stereochemistry of nimbin. *Tetrahedron*, **24**, 1517–1523.

Hegnauer, R. (1969) *Chematoxonomie der Pflanzen*. Vol. 5, Birkhasuer Verlag, pp. 64–71.

Henderson, R., McCrindle, R., Malera, A. and Overgon, K.H. (1968) Tetranortriterpenoids IX. The constitute and stereochemistry of salannin. *Tetrahedron*, **24**, 1525–1528.

Katani, R. and Padhya, M.A. (1988) Isolation of nimbin from Azadirachta indica and its callus culture. *Indian Drugs*, **25**, 526–527.

Ketkar, C.M. and Ketkar, M.S. (1993) Azadirachtin content of neem and its by products from different parts of India (abstr.) *Proc. World Neem Conf.*, Bangalore, India, 24–28 February, 1993.

Kraus, W. (1983) Biologically active compounds from meliaceae. *Proced. 2nd Inter. Conf. Chem. Biochem. of Active Natural Compounds*, Budapest, Hungary, pp. 331–345.

Kraus, W. (1986) Constituents of neem and related species. A revised structure of azadirachtin. In W. Atta ur-Rehman and P. Le Quesne (eds.) *New Trends in Natural Product Chemistry*. Elsevier, NY, pp. 237–256.

Kraus, W., Bokel, M., Bruhn, A., Cramer, R., Klaiber, I., Klenk, A., Nagi, G., Pohnl, H., Sodlo, H. and Vogler, B. (1987) Structure determination by NMR of azadirachtin and related compounds from Azadirachta indica A. Juss (Meliaceae). *Tetrahedron*, **43**, 2817–2830.

Kraus, W., Bokel, M., Klenk, A. and Pohnl, H. (1985) The structure of azadirachtin and 22, 23 dihydro 23-methoxyazadirachtin. *Tetrahedron letter*, **26**, 6435–6438.

Kobeleswaran, V., Rajan, S.S., Govindachari, T.T. and Gopalkrishna, G. (1994) Crystal and molecular structure of azadirachtin A. *Current Science*, **55**, 362–364.

Koul, O., Isman, M.B. and Ketkar, C.M. (1990) Properties and uses of neem (Azadirachta indica). *Canadian Journal of Botany*, **68**, 1–11.

Lavie, D., Levy, E.C. and Jain, M.K. (1971) Limonoids of biogenetic interest from Melia azadirachta. *Tetrahedron*, **27**, 3927–3939.

Ley, S.N. (1994) Synthesis and chemistry of insect anti-feedant azadirachtin. *Pure and Applied Chemistry*, **66**, 2099–2102.

Mitra, C.R. (1963) *Neem*. Indian Central Oilseed Committee, Hyderabad (India).

Mordue, A.J. and Blakewell, A. (1993) Azadirachtin—an update. *Journal of Insect Physiology*, **39**, 903–924.

Morgan, D.D. and Mandava, N.B. (1987) *Handbook of Natural Pesticides*, Vol. III, *Insect Growth Regulators*, Part B., CRC Press Inc., Florida, USA.

Nat, J.M. van der, Sluis, W.G. van der, De Silva, K.T.D. and Labadie, R.P. (1991) Ethnopharmacognostical survey of Azadirachta indica A. Juss. (Meliaceae). *Journal of Ethnopharmacology*, **35**, 1–24.

Pant, N., Garg, H.S., Madhusudanam, K.P. and Bhakuni, D.S. (1988) Sulfurous compounds from Azadirachta indica leaves. *Fitoterapia*, **57**, 302–304.

Pattabiraman, T.N. (1978) Studies on plant gums. Part III Isolation and characterisation of a glycoprotein from neem (Azadirachta indica). *Indian Journal of Biochemistry and Biophysics*, **15**, 449–455.

Pattabiraman, T.N. and Lakshmi, S.U. (1968) Amino acid protein in plant gum. *Science and Culture* (Calcutta) **34**, 68–70.

Rangaswamy, S. and Parmar, B.S. (1994) Azadirachtin content at different states of flowering and fruiting in neem. *Pesticide Research Journal*, **6**, 193–194.

Rastogi, R.P. and Mehrotra, B.N. (1990) *Compendium of Indian Medicinal Plants*. Vol. I, Publication and Information Directorate, CSIR, New Delhi, India, pp. 50–52.

Rastogi, R.P. and Mehrotra, B.N. (1991) *Compendium of Indian Medicinal Plants*. Vol. II, Publication and Information Directorate, CSIR, New Delhi, India, pp. 87–90.

Rastogi, R.P. and Mehrotra, B.N. (1995) *Compendium of Indian Medicinal Plants*. Vol. III, Publication and Information Directorate, CSIR, New Delhi, India, pp. 85–88.

Remboldt, H. (1989) Isomeric azadirachtin and their mode of action. In M. Jacobson (ed.) *Focus on Phytochemical Pesticide*, Vol. I, *The Neem Tree*, CRC Press Inc., Florida, USA, pp. 47–67.

Remboldt, H., Forster, H. and Czoppelt, C. (1987a) Structure and biological activity of azadirachtin A and B. In H. Schmutterer and K.R.S. Ascher (eds.) Natural Pesticides from neem tree Azadirachta indica, and other tropical plants. *Proc. 3rd Inter. Neem Conf.*, Nairobi, Kenya, 10–15 July, 1986.

Remboldt, H., Forster, H., Sonnenbichler; J. and Klenk, A. (1987b) Structure of azadirachtin B. *Zeitschrift fur Naturofrschung, C-Biosciences*, **42**, 4–6.

Remboldt, H., Puhlmann, I., Balan, K. and Kumar, C.S.S.R. (1993) Phytochemistry and structure activity relationship of azadirachtin. *Proc. World Neem Conf.*, Bangalore, India, 24–28 Feb. 1993.

Rengasamy, N., Kaushik, N., Kumar, J., Koul, O. and Parmar, B.S. (1993) Azadirachtin content and biochemistry of some ecotypes of India. (abstr.) *Proc. World Neem Conf.*, Bangalore, India, 24–28 February 1993,

Schmutterer, H. (1995) *Neem Tree, source of unique natural product for integrated pest management and medicinal, industrial and the other properties*. VCH Verlagschaft Weinhim, Germany.

Schneider, B.H. and Ermel, K. (1987) Quantitative determination of azadirachtin from neem seeds using high performance liquid chromatography. In H. Schmutterer and K.R.S. Ascher (eds.) Natural pesticides from neem tree (Azadirachta indica) and other tropical plants. *Proc. of 3rd Inter. Neem Conf.*, Nairobi, Kenya, 10–15 July, 1986.

Schroeder, D.R. and Nakanishi, K. (1987) A simplified isolation procedure for azadirachtin. *Journal of Natural Products*, **50**, 241–244.

Sen, R.N. and Banerjee, G. (1931) The bitter principle of neem oil. *Journal Indian Chemical Society*, **8**, 773–776.

Siddiqui, S. (1942) A note on the isolation of three new bitter principles from the neem oil. *Current Science*, **11**, 278.

Siddiqui, S. and Mitra, C. (1945a) Isolation of bitter principles from neem oil. *Indian Patent* No. **33649**, 26 Nov. 1945, 15 Nov. 1946.

Siddiqui, S. and Mitra, C. (1945b) Separation of the bitter principles of neem oil and simultaneous refining of the oil. *Indian Patent* No. **33650**, 26 Nov. 1945, 15 Nov. 1946.

Siddiqui, S. and Mitra, C. (1945c) Manufacture of new derivatives of subsidiary products, the physiologically active bitter principles of neem oil. *Indian patent* No. **33651**. 26 Nov. 1945, 16 Nov. 1946.

Siddiqui, S., Siddique, B.S., Faizi, S. and Mahmood, T. (1988) Tetracyclic triterpenoids and their derivatives from Azadirachta indica. *Journal Natural Products (LJyodia)*, **57**, 30–43.

Sidu, O.P. and Behl, H.M. (1996) Seasonal variation in azadirachtin in seed of Azadirachta indica. *Current Science*, **70**, 1086.

Sinha, N.P. and Gulati, K.C. (1968) Amino acid content of neem seed cake. *Proceedings National Academy of Sciences* (India), **38**(A), 151–154.

Singhal, K.K. and Mudgal, V.D. (1984) Studies on chemical composition of neem cake and release of ammonia from neem coated urea. *Indian Journal of Dairy Science*, **37**, 285–287.

Skellon, J.H., Thorburn, S., Spence, J. and Chatterjee, S.N. (1962) The fatty acids of neem oil and thir reduction products. *Journal Science Food Agriculture*, **13**, 639–643.

Subramanian, S.S. and Rangaswamy, S. (1947) Chemical examination of the flowers of Melia azadirachta. *Current Science*, **16**, 182–183.

Subramanian, M.S. and Lakshmanan, K.K. (1993) Azadirachta indica A. Juss. Neem drugs, as an antileprosy source. *Proc. World Neem Conf.*, Bangalore, India, 24–28 Feb. 1993.

Sundaram, K.M.S. and Curry, J. (1993) High performance liquid chromatographic determination of Azadirachtin in conifers and deciduous foliage, forest soil, leaf litter and stream water. *Journal of Liquid Chromatography*, **16**, 3275–90.

Tewari, D.N. (1992) *Monograph on Neem.* International Book Distributors, Dehradun (India).

Thakur, M.S. and Godrej, B.P.C. (1972) Purifying neem oil. *Indian Patent*, **118678**, 10 Oct. 1970.

Tirimanne, A.S.L. (1984) Surveying the chemical constituents of the neem leaf by two dimensional thin layer chromatography. *Proc. of the 2nd Inter. Neem Conf. Germany.*

Vashi, I.G. and Patel, H.C. (1988) Amino acid content and microbial activity of Azadirachta indica A. Juss. *Journal Institute of Chemists*, **60** (Part II), 43–44.

Warthen, J.D. Jr., Stokes, J.B., Jacobson, M. and Kozempel, M.F. (1984) Estimation of azadirachtin content in neem extract and formulations. *Journal of Liquid Chromatography*, **7**, 591–598.

Wealth of India (1948) Publication and Information Directorate, CSIR, New Delhi (India).

Yakkundi, S.R., Tejanathi, R. and Ravendranath, B. (1995) Variation of Azadirachta content during growth and storage of neem Azadirachta indica seed. *Journal of Agriculture and Food Chemistry*, **43**, 2517–2519.

Yamasaki, R.B. and Klocke, J.A. (1987) Structure bioactivity relationships of azadirachtin—a potential insect control agent. *Journal Agriculture and Food Chemistry*, **34**, 467–471.

Yamasaki, R.B., Lee, S.M., Stone, G.A. and Darlington, M.V. (1986) Isolation and purification of azadirachtin from neem (Azadirachta indica) seed using flash chromatography and High performance liquid chromatography. *Journal of Chromatography*, **365**, 220–226.

Yamasaki, R.B., Ritland, T.G., Barnby, M.A. and Klocke, J.A. (1988) Isolation and purification of salannin from neem seed, and its quantification in neem and chinaberry seed and leaves. *Journal of Chromatography*, **447**, 227–283.

4. CULTIVATION

INTRODUCTION

Soil in many parts of the world has been degraded because of over-exploitation by man and deforestation. Neem has been found to be a tree of choice for afforestation. It has been found suitable for arid regions and for large-scale cultivation (Radwansky and Wickens, 1981; Chaturvedi, 1985) and the agroforestry of neem has been given a new term, "Margoculture". Cultivation of neem has been recommended for the following reasons:

1. Multifarious uses: leaves for fodder (Nehra *et al.*, 1987), wood for timber and fuel, and seed for oil, manure and pesticide.
2. Tolerance of tree to high temperatures, aridity and high concentration of salts in the soil.
3. Avenue tree, which also acts as a wind-breaker (Oboho and Nwoboshi, 1991), shelter belt, canal-side plantation and sand dune stabilizer.
4. Leaf litter enriches the soil by providing organic matter without any harm to friendly insects, earthworms etc.
5. It adapts itself very well to degraded and acid soil and can be used as a tree for reclamation of wasteland (Sastry and Kanathekar, 1990).

Because of the above, margoculture has been included in agroforestry projects in most of the developing countries of the world with dry and arid zones. In developed countries, wherever possible, it is being cultivated as a raw material for harmless pesticides which can be obtained from the seed. Scientific studies are being conducted for methods of mass propagation, ideal conditions for growth and development, and also for its impact on the eco-balance, environment and agricultural crops. Werner and Muller (1990) described it as a fast-growing tree. Ketkar (1982) gave the preparation and uses of its products and by-products. Randhawa and Parmar (1993) gave an account of research on neem.

METHODS OF CULTIVATION

From Seed

Neem seed, in general, loses its germination capacity after a couple of weeks of maturity. Efforts have been made to find out the reasons for this small period of viability and how the germination period of seed can be extended by modifying storage conditions (Jha and Chaudhuri, 1995).

Reasons for Loss of Germination Capacity

Various views have been expressed; some of the important ones are:

1. Fermentation

It has been postulated that in the kernel enclosed in the endocarp, a breakdown of sulfur-containing compounds takes place, causing damage to seed. A viable seed has a green cotyledon, but it turns brown after a couple of days in hot and humid conditions (Smith, 1939). The dry viable seed, when immersed in water, emits a garlic-like smell, which is not present in non-viable seed, indicating loss of some sulfur-containing compounds.

2. High Temperature

Seed stored at low temperature remain viable for a longer period, indicating thermal sensitivity of the embryo. It has been postulated that high temperature increases the metabolic rate of the seed embryo, causing its death.

Seed of the same batch, stored in India at an average room temperature of 30°C with low humidity and in Florida (USA) at an average room temperature of 20°C with high humidity, when studied after couple of months were found to have a green cotyledon in most of the seed in Florida, but the seed cotyledons had turned brown in India, with most of the seed full of fungal spores.

3. Due to Pulp

Seed without pulp give a much better germination percentage, as compared to those with pulp (Prasad, 1941). The fruit swallowed by birds, for its sweetish mucilage, gets de-pulped in the alimentary canal of the birds and, when dropped, these seed have a higher germination rate. It has been postulated that the germination percentage in seed with pulp decreases because of the meager amount of moisture present. This amount of moisture is sufficient to start the early stages of embryo development but cannot sustain it until completion of germination, thus causing death of the embryo in the early stages, resulting in loss of germination capacity.

4. Endocarp

It has been known to form a water- and air-tight compartment. If the seed are dried at high temperature, as has been observed by Puri and Hardman (1980) in *Trigonella foenum graecum*, they become dormant and are said to be "hard". These seed fail to germinate until the endocarp is punctured by some physical or chemical means. A similar phenomenon may be operating in this case also. It has been seen that in neem, in those cases where the endocarp was removed, the seed showed a better germination.

5. Inhibitory Compounds

These may be present in the seed coat or in pulp or in the endocarp and may not allow the seed to germinate (Ponnuswamy *et al.*, 1993).

6. Biochemical and Physiological Barriers

There may be some peculiar mechanism operating in the embryo which does not allow further development after a fixed period of time, causing loss of germination capacity.

Storage and Other Conditions for Germination

Chaisurisri *et al.* (1986) dried seed at 25°C in an air-conditioned room and in a glasshouse under other conditions. The most effective treatment was drying the fresh seed in sunlight for 3 days at 46.16 percent moisture before storing them in cotton bags at 15°C. The seed stored in this way retained their viability for more than four months, giving 62 percent germination. Nagaveni *et al.* (1987) collected greenish yellow seed, when they were not fully ripe, de-pulped them and dried them in shade for 2 days. A germination rate of up to 80 percent could be achieved from these, even after four months. After six to seven months germination decreased to 50 percent. Maithani *et al.* (1989) studied the effect of fruit maturity, temperature and the container on germination. The maximum germination was found from seed 10–12 weeks old after flowering, when fruit turned yellow. Seed stored at 15°C or 50°C in sealed or perforated plastic bags, cardboard boxes, etc., deteriorated, while those stored in aerated containers or at 15°C, showed about 15 percent germination after six months. Venkatesh *et al.* (1990) also obtained good germination by storing de-pulped seed in cotton bags for up to four months, but after that there was a decline.

Chaney and Knudson (1988) and Radhamani *et al.* (1990) found that removal of the endocarp improves germination considerably. The authors postulated that the endocarp develops a physical barrier for water, gases, enzymes and inhibitors and for metabolism of fats.

Ponnuswamy *et al.* (1991) graded the de-pulped seed on the basis of their behavior when immersed in water. Those remaining on the surface were labeled "floater" and the others "sinker". The floaters were 18 percent of the total and had low viability as compared to sinkers which had 90 percent germination. After 3 months of storage, the germination rate dropped to 15 percent. Seed stored in earthen pots and buried in moist sand recorded 62 percent germination even after 3 months of storage.

Aflatoxin may be one of the causes for lowering the quality of seed when stored in an enclosed atmosphere at high temperature and humidity. Chourasia and Roy (1991) studied the effect of temperature, light and humidity on the storage of neem seed and the production of aflatoxin B1. The highest production of this compound was recorded at 30°C and relative humidity of 96 percent.

Nursery Techniques

Seed after harvesting should be immediately de-pulped, washed with water and dried in shade with a free flow of water, to such an extent that the outer surface of the seed coat should not be wet to the touch. These seed can be sown in nursery beds, about 2.5 cm deep and at a distance of 2.5 cm each. The soil should be loose and sprinkled with water, but there should be no water logging. Seed can be sown in small containers (flower pots) or in thick plastic bags filled with soil, at the rate of 2–3 seed per container. In the majority of cases, the seed germinate within couple of days, but some seed may take a longer time. When the seedlings are about 2 cm high, they can be transplanted from the nursery beds to bigger beds at a distance of 15 cm × 15 cm. In cases where more than one seed is sown in pots/bags, the most robust seedling should be retained, the rest pulled out and discarded.

According to Vijayalakshmi *et al.* (1995), raised beds of 10 × 1 m in area and 15 cm in height should be prepared. Farmyard manure, sand and local soil should be mixed in the ratio of 1:1:3. This mixture should be put on top of soil to a height of 2.5–5 cm. For transplanting, the bag should be filled with soil containing silt, sand, clay and farmyard manure in the ratio of 1:1:1:1. The plastic bags should be of 150–200 gauge thick so that they have enough mechanical strength and they do not burst during shipping.

In areas where there is a possibility of attack by insects on young seedlings or by ants and termites and the environmental conditions are not ideal, the nursery seedlings can be raised in trays, placed on a platform in a greenhouse (Figure 8). A greenhouse can be economically erected by making a frame of water pipes covered over with thick transparent plastic sheet. On one side exhaust fans or a big blower should be installed so that there is free movement of air. Water can be sprinkled with sprinkler-attached pipes inside the greenhouse or by hand sprinkler. This type of improvised greenhouse is

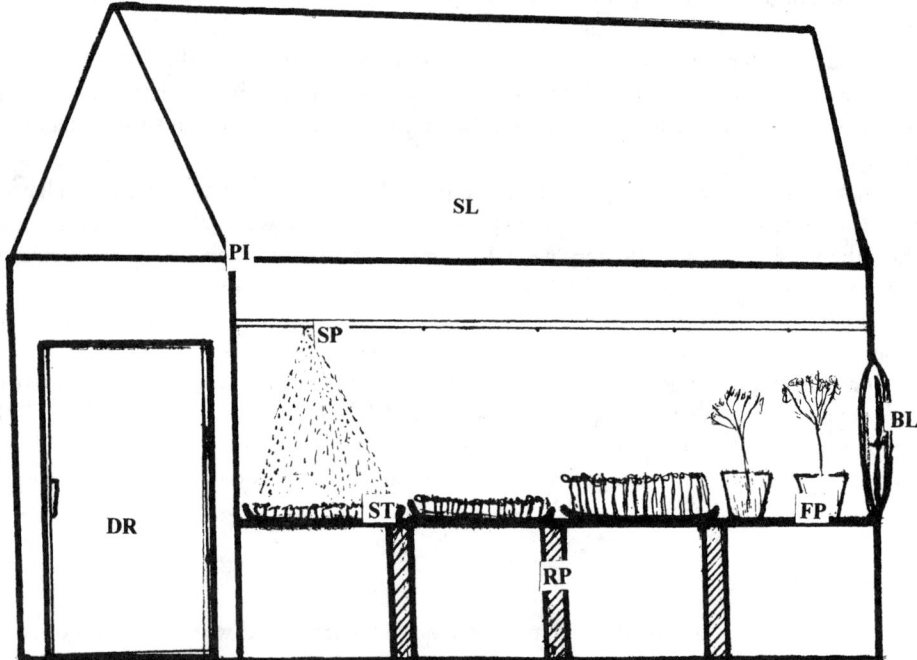

Figure 8 A simple green house for young seedlings *Abbreviations*: BL = blower, DR = door, FP = flower pot, PI = pipes for supporting the structure, SL = transparent plastic sheet to cover the green house, SP = sprinkler, ST = seed tray, RP = concrete stand

very useful in areas with high humidity, heavy rainfall and frost. Water mist has to be created inside, otherwise the seedlings get dehydrated by the greenhouse effect, which is evident by the development of pink pigment in the leaves and hypocotyl, and the stunted growth of the seedlings.

Further details about nursery techniques can be obtained from Maithani et al. (1988), who have given the method of sowing neem in moist tropical climatic conditions. Joshi and Prakash (1992) studied extracts of some trees on germination and seedling growth. The seedlings are ready for transplanting in pits when 6 months old and 15–25 cm in height. The pits of $30 \times 30 \times 30$ cm should be dug at intervals of 3×3 m, preferably before rain, and watered frequently.

Direct Sowing

For degraded areas, direct sowing has been recommended by Tewari (1992), provided there is protection and enough moisture during the initial stages. This may be done by dibbling in bushes, by making small holes in a bush like *Euphorbia*, and then sowing 2–3 seed and sealing the hole. Broadcast sowing can be done on ploughed or unploughed soil, while sowing in lines can be done along the ploughed fields of crops. In heavy soils, sowing on mounds or ridges was practiced, whereas in dry sites trenches were preferred for retaining the moisture. Contour ditches have also been tried in some areas (Nagaraju Kumar et al., 1992).

Vegetative Propagation

Neem seed has a very small period of viability and various methods have been developed to prolong it, which have been discussed earlier. Bhardwaj and Gurdev Chand (1995) suggested cryogenic cold storage. Even if seed germination is achieved, a plantation sown from seed may have a lot of variability. Vegetative propagation may be helpful because of early maturity, true-to-type stock and uniformity of the growth characteristics of the plantation. Good mother trees can be selected on the basis of the desired characteristics such as good vigor, disease resistance, age, better crown shape, well distributed branches, and big, abundant seed rich in azadirachtin. Jha and Chaudhari (1990) carried out trials on stump planting and Swaminathan and Surendram (1989) carried out studies on rooting response to growth regulators. For rapid multiplication of the elite tree, Mohinderpal (1995) prescribed the following methods:

1. *Grafting and budding*: Early summer is the best season. Wedge grafting and patch budding techniques gave good results. Pong-Anant et al. (1989) gave further details.
2. *Air layering*: One cm wide bark was cut by Shanmungavelu (1967) and the exposed tissues were treated with indole butyric acid (IBA) or 1 naphthalene acetic acid (NAA) and covered over by moist moss or coconut fibers. Only those air layers that were treated with 0.1 percent NAA and IBA produced roots.
3. *Rooting branch cuttings*: Hardwood, semi-hardwood, softwood and juvenile cuttings are used:
   (a) *Juvenile root* suckers were planted in a moist layer of sandy soil (Rao, 1958; Pal et al., 1994).

(b) *Hardwood*: Cuttings were treated with IBA and planted in a high humidity area. Sivaganam *et al.*, (1989) observed that these cuttings rooted under mist in 135 days with 1000 ppm of IAA and IBA. The treatment, given by a drip method, stimulated rootings.
(c) *Semi-hardwood*: Cuttings treated as above but for a short duration.
(d) *Softwood cuttings*: In this case, long binodal leafy cuttings are taken. Pal *et al.* (1992) stimulated the root response by treatment with phenolic compounds like naphthol and salicyclic acid, under mist. Verma *et al.* (1996) experimented with the effect of auxin on rootings of cuttings in the spring season by applying IBA and NAA. The 100 ppm of both the hormones significantly increased rooting and sprouting but IBA was more effective. Chander *et al.* (1996) studied the effect of auxins on the number of leaves per cutting and the length/age of the main branch. The cuttings, taken from a 8–10-year-old tree and treated with 500 ppm IBA and 2000 ppm of IAA, resulted in maximum increase in the number of branches and leaves.

BIOTECHNOLOGY AND TISSUE CULTURE

Tissue culture techniques are an excellent tool in neem research to obtain organogenesis and plant regeneration in a short time under controlled conditions. By using these techniques, we are not dependent on the germinability of the seed, which is of very short duration, and will be able to obtain pesticidal constituents which could not be synthesized so far (Tewari, 1992), or whose synthesis may not be economically viable.

Work on this aspect has been carried out for the past several years. Sanyal *et al.* (1981) studied *in vitro* hormone-induced chemical and histological differentiation in stem culture, while Ekundayo (1983) gave an account of the biosynthesis of nimbolides from (2–14C) mevalonate and (2–14C) acetate. Schulz (1984) obtained callus proliferation from various parts of the plant by improving the culture media. Pratap and Jaiswal (1985) tried plantlet regeneration for leaflet callus from 15-year-old trees. Data were collected on percentage root and shoot production under all treatments. Root formation from calli occurred wherever 1 naphthalene acetic acid (NAA) was used, whereas addition of a low concentration of NAA to a medium containing benzyladenine increased the frequency of shoot formation. The callus-derived shoots produced roots and developed into plantlets. Sarkar and Datta (1986) established the relationship between biosynthesis of nimbin and β sitosterol in bark and bark-originated callus of increasing age and studied the effect of glycine on *in vitro* biosynthesis of nimbin and β sitosterol in tissues. Sanyal *et al.* (1988) found that the cotyledons contained nimbin, glycine, other amino acids and β sitosterol. The effect of glycine on the synthesis of nimbin and β sitosterol was investigated and it was observed that glycine affected the synthesis of both. Ramesh Kumar and Padhya (1988) isolated nimbin from the leaves and its callus culture growth. Indole acetic acid (IAA) and indole butyric acid (IBA) interaction showed a linear increase in nimbin content. Rao *et al.* (1988) also studied the growth of callus. For genetic transformation, Naina *et al.* (1989) induced tumors in the culture using *Agrobacterium tumefaciens*. The transformed plantlets were regenerated from the infected region of the seedling and also from tumor cells.

Ramesh Kumar and Padhya (1990) propagated plantlets from leaf disks. These workers (Ramesh Kumar and Padhya, 1993) worked further on this aspect and obtained 18–20 adventitious shoot buds within four weeks by variously treating epidermal cells, which gave rise to 12–15 plantlets within six months. These were later transferred to the plots, where stable lines were developed from them.

Gautam et al. (1993) studied the development of root and shoot in anther-derived callus. The callus originated in the microsporangial wall layer and connective tissue of anthers. After 13–15 weeks, a green nodular structure and prominent roots developed. A different culture medium induced multiple shoots. Nirmala Kumari et al. (1993) observed that the growth physiology of the basal region of hypocotyl appears to be fundamentally different from other shoot regions and holds good potential for multiplication through cuttings of seedlings and tissue culture propagation. The various explants differentiated for embryoid formation and produced leafy shoots. Stephen et al. (1993) produced callus without further differentiation from basal hypocotyl segments. Similar callus at the cut surface of the cotyledons produced roots. In a one-month-old basal hypocotyl segment, several lenticels were produced which gave rise to shoot buds. Joarder et al. (1993) cultured nucellus tissue of immature seed after fertilization. Five to seven plants were regenerated per culture. Immature cotyledons of the age group of 60–70 days were found to be the best for embryogenesis, but the majority of the cultures produced embryoids which did not germinate. Drew (1993) developed a technique for micro cuttings *in vitro* for clonal propagation of neem lines with a high yield of azadirachtin and free of genetic off-types.

According to Thiagarajan and Murali (1994), the optimum conditions for high frequency regeneration of plantlets can be obtained from a medium containing 3 percent sucrose supplemented with 0.1 mg/L NAA and 0.3 mg/L of 6-benzoylamino purine. The regenerated plants were successfully established outdoors. Allan et al. (1994) cultured embryos from fruits, collected thirty days after pollination. The culture yielded 0.0007 percent of azadirachtin on a dry weight basis. Newton et al. (1995) described somatic embryo genesis, embryogenic culture and culture maintenance in neem along with other woody plants.

Peroxidases isoenzyme of organogenesis were used by Preetha et al. (1995) as biochemical markers for differentiating regenerative callus lines in the early stages of differentiation. These authors gave the methods of callus induction, sample preparation and electrophoresis. The isoenzyme characters of the various types of cultures were discussed in detail. Clonal propagation of neem from non-woody explants was achieved by Joshi and Thengane (1996) by supplementing an MS medium with benzyladenine and kinetin. Two to three shoots per explant were observed after twenty days. The multiplication frequency of shoots per explant and the average shoot length were observed after 14–16 weeks. Rootings from these shoots developed, and regenerated plantlets were established in the soil.

GENETIC IMPROVEMENT

Neem is a heterogeneous group showing a great amplitude of variation. The wide distribution of neem and its growth under a variety of climatic conditions amply demonstrate the genetic diversity present within the species, which can be utilized for genetic improvement, but traditional methods of breeding, including hybridization, can be of very limited use owing to their long generation time and the prevalence of outbreeding. The other methods which can be utilized for the above purpose, as suggested by Tewari (1992), are:

1. Selection of plus tree from variations within the wide population. The international neem network has also recommended investigation on the evaluation of genetic variability through provenance trials. A number of studies have been conducted in India in this direction.
2. Germplasm bank.
3. Seed production for mass multiplication for a selected superior clone.
4. Mutation breeding.
5. Polyploid breeding.
6. Modern techniques: biotechnology, genetic engineering, tissue culture, callus formation, somatic embryogenesis, direct multiple shoot induction from seedling explant, genetic transformation by *Agrobacterium tumefaciens*, which infects a large number of plant species and causes tumor formation, etc. may be utilized.

Chinnamani (1993) and Read and French (1993) suggested a strategy for neem improvement.

DISTRIBUTION IN INDIA

In the literature it is often repeated that this tree is distributed all over the Indian subcontinent, but it is a general statement. The tree does not prefer silty, saline, mica-rich or inundated soil. It cannot bear frost or high humidity or water-logged conditions. It is light demanding and is often cultivated near human habitation for its multifarious uses and also as an avenue or roadside tree.

Champion and Seth (1968) mentioned its occurrence in tropical dry deciduous thorn forest and in tropical dry evergreen forest as follows:

- Southern tropical dry deciduous forest
- Southern tropical thorn forest
- Tropical dry evergreen forest
- Northern tropical thorn forest

DISTRIBUTION ALL OVER THE WORLD

The naturalization of *A. indica* in the countries adjacent to the Indian subcontinent, which include Thailand, some parts of Malaysia, Java and Bali (now Indonesia) and probably east Africa, can be explained on the basis of the long cultural and commercial relations that these areas have had with India since ancient times.

In Asia, in addition to the Indian subcontinent the tree has been reported from upper Myanmar, drier parts of Sri Lanka (earlier known as Ceylon), south east Thailand, Vietnam, southern Malaysia, drier islands east of Java in Indonesia, the Philippines, Hainan in China, middle east Yemen, Qatar, and some parts of Saudi Arabia.

In the past two centuries it was introduced into other parts of the world by immigrants from India who settled in east Africa, the Caribbean islands, Fiji, Mauritius etc. It also grows in the islands of the South Pacific, the West Indies, Haiti, Surinam, the Dominican Republic, Cuba, Nicaragua, and in some areas of Mexico. It was introduced into selected pockets of California, south Florida (USA) and Queensland (Australia).

It adapted well to the savannah conditions south of the Sahara in Africa, so introduction has been very fast in some of the African countries. It grows well in Egypt, Sudan, Ethiopia, Somalia, Kenya, Uganda, Tanzania, Mozambique and Chad in the northern and eastern part of Africa and in Mauritania, Senegal, Mali, Ghana, Ivory Coast, Togo, Nigeria and Cameroon in the rest of Africa.

Recent Introduction in Various Geographical Regions

Cultivation from Haiti was reported by Lewis and Elvin-Lewis (1983), who recommended the large-scale introduction of this tree in other neotropical countries, as the tree tolerated the arid conditions. It was cultivated on a large scale in Upper Volta (Sieder, 1983). Ujhmura (1986) presented a case study of neem introduction in semi-arid zones of Nigeria.

Ahmed *et al.* (1989) reported an establishment of 50,000 trees in the plains of Arafat in Saudi Arabia. Fallata people who originated from West Africa and now live in Sudan have long experience of planting neem trees (Kismul, 1989). Silvicultural characteristics and a manual for the propagation and afforestation in Sudan have been given by Badi *et al.* (1989) and Branney (1989), while Oguntala (1989), after studying the wild forest fires in Nigeria, found that the survival of exotic species like neem after fire was higher as compared to indigenous ones. In the rain-fed zones of Indonesia, Tampabolon and Alrasyid (1989) discussed the prospects of neem in the rehabilitation of critical land. Spaak (1990) included neem as one of the trees in the afforestation program on Cape Verde in an arid and semi-arid zone which has extensively high annual rainfall. Adegbehin *et al.* (1990) gave an account of an agroforestry project near the shelter belt in the Savannah area of Nigeria.

In the northernmost province of Cameroon, 25 to 30 percent of the land is composed of sterile or degraded soil. In an attempt to rehabilitate this soil, neem, along with other trees, were planted by Matig (1989), and the results were very encouraging. There was spontaneous regeneration of the natural vegetation. On the west coast of Reunion, Roederer (1991) gave an account of forestry and agroforestry experiments with neem. The tree survived well but grew very slowly. In Chad also, neem was introduced to improve the forest resources (Thomossey, 1991). Chiu (1993) reported successful introduction in the Hainan Province of China.

Dry Zone Afforestation

Encouraged by the survival of the tree in near-drought conditions, experiments were conducted on the introduction of neem in areas having degraded and arid soil. For degraded forest areas, direct sowing was more successful and cheaper, provided that protection and shelter were available to young seedlings. Direct sowing was by dibbling in bushes, broadcasting in lines on mounds and ridges, in trenches, and in sunken beds. Aerial seeding gave profuse germination on soil in Nigeria but all seedlings in direct sunlight died and only those in the shade of natural shrubs could survive during the dry season. Kalla *et al.* (1978) gave an account of the felling cycle of trees in the Indian desert for integration of forage forestry in an afforestation program. Neem, along with other trees, was found suitable for the degraded condition of forest areas in low rainfall regions by Raddi (1981). In the Indian desert three shelter beds were planted in 1973 by Muthana *et al.* (1984). The belts had three rows of trees each and they were laid out perpendicular to the wind direction. The outside rows contained *Acacia tortilis*, *Prosopis juliflora* and *Cassia siamea*, while the center rows were planted with *A. indica*, *Albizzia lebbeck* and *Eucalyptus camaldulensis*. The shelter beds were easily established and grew well. Ahmed *et al.* (1985), on the other hand, planted neem trees in coastal sand, using highly saline water for irrigation. An optimum spacing of 140 m for neem in shelter bed establishment was recommended by Onyewotu (1985) in the Sahel and Sudan zones of Nigeria. Bose and Bandoyopadhyay (1986), during a study of the economics of energy plantation in alkaline soil near Delhi (India), found a mortality rate in neem quite low after 3 months of planting, but height was least as compared to the other trees. These authors felt that fuel wood farming in this area was economically feasible. Chaturvedi (1985), in a forest bulletin, gave an account of the firewood farming of 16 important trees on degraded lands in the Gangetic plains in India. Neem was also included in the list. The author discussed the ecology of sites on saline, sodic and alkaline *user* (infertile) soils, ravines and areas of brackish water, with growth and yield statistics. The salt tolerance of neem tree seedlings was studied by Gupta *et al.* (1987) by carrying out greenhouse experiments on sand, loam and clay forest soils to which calcium chloride had been added. Toxicity was lowest in the clay soils, and greatest in sandy soils.

An effective method of wasteland afforestation was developed by Kinhal (1968) by early planting and critical watering, in pits at a spacing of 3×2 m. Martinez (1987) gave an account of the silviculture of neem under multi-purpose tree. Panchan *et al.* (1989) studied the effect of salinity on the growth of the neem tree, by applying different concentrations of salt (sodium chloride). Neem did not tolerate even 0.4 percent salt. Oguntala (1989) studied a forest fire climate in relation to fire incidents. After the fire, the survival and regeneration in neem was higher as compared to the indigenous species. A technique was developed by Mehta (1989) for planting trees on highly alkaline soil, by levelling and bunding the soil and applying gypsum over the whole surface, followed by the cultivation of crops for 4–5 years. The trees were planted afterwards.

Prasad and Dhuria (1989), for the reclamation areas mined for iron ore, carried out afforestation trials with neem by planting it in pits. Half of the pit was filled with original soil and the other half with humus-rich soil and manure. Biomass data was collected. Various anti-desertification techniques were suggested by Campolucci and Paolim (1990). These suggestions were (a) contour terracing, (b) half-moon-shaped *bunds*, (small dams), (c) furrows up to 80 cm deep and (d) deep ploughing, followed by two central rows of container-grown trees of *Acacia* spp. and *Azadirachta indica*, to act as a wind-breakers; flanking them were two rows of other trees and six rows of shrubs to act as a protective barrier against browsing damage.

Desh Raj (1990) developed land which was without any vegetation but had large patches of white salts. Neem did not grow well on this soil.

Mohan et al. (1990) observed that neem grew best in clay soil. To confirm this, the authors used various potting mixtures by using different doses of soil and fertilizers. The mixture of sand–clay–farmyard manure with 30 ppm nitrogen and 20 ppm phosphorus was found to be the best. Singh et al. (1991) gave uses of various soil-working techniques as follows: (a) pit planting, (b) ploughing to 20 cm depth, and (c) digging V-shaped furrows 30 cm deep. The ploughed areas gave the best results, while Ahmed and Puzari (1991), in a growth data trial, got their best results from neem trees at a livestock research station. The tree did not fare well in coastal areas. Gill and Abrol (1991) gave a summary of the results of field experiments carried out since 1971 at the Central Soil Salinity Research Institute in India. The studies showed that the alkali soil of the Indo-Gangetic plains of north India was toxic to neem because of an excess of sodium carbonate and bicarbonate, poor soil structure and moisture transmission. The soil mixed with gypsum and farmyard manure supported some trees, but neem did not show high tolerance to these conditions. The neem tree was found superior in enriching the sandy, loamy soil with calcium and in increasing soil pH by measuring litter quality and soil fertility (Drechsel et al., 1991). Verinumbe (1991) developed an agroforestry system for neem and the other trees on vertisol soil. In a wasteland reclamation project, neem was planted with other trees as a fuel tree by Chowdhury (1992). The project was a success. The survival rate of neem was 83.33 percent after 3 years. In other experiments, neem trees were planted in arid regions (Harsh et al., 1992) for control of soil erosion and fertility enhancement of the soil (Laskar and Datta, 1992). To find the effect of the neem tree on the yield of wheat in arid zones, neem trees were planted in the wheat field by Puri et al. (1995). The authors did not observe any significant difference in the yield of farm grain in the fields with and without neem trees. Dalal et al. (1992) also worked on this direction.

Shaikh (1992) suggested that in highly eroded and degraded soil, seedlings should be planted in pits with shoulder trenches, and that for water harvesting, micro catchment areas should be formed. Shaikh (1993) suggested a change in strategy for neem tree plantation in highly eroded and degraded areas. Neem was not found suitable for sodic soil aquic petrocalcic natrustalf soil having pH 9.7 (Sharma et al., 1993). Harikrishanan (1993) gave an account of nine species, including neem, for small marginal farmers.

The rooting pattern was studied by Toky and Bist (1992), with further research on the growth pattern and architectural analysis of neem along with other trees (Bist and Toky, 1993) in a six-year-old plantation. These authors observed that the period of growth, which began in the dry season, was completed in the rainy season that followed. Brenner et al. (1995) studied wind-break crop interaction behind two rows of neem trees in a field of millet.

In arid areas, proper moisture for the root system is essential, particularly at the initial stages of tree growth. Gupta (1992) studied the growth of nursery plants of neem as influenced by soil mixture and fertilizers. Meena et al. (1995) discussed various soil techniques on early growth in arid zones, while Gupta (1995) studied different systems/combinations of rain/water harvesting. The ridge and furrow method was found to be the best and significantly improved the growth of trees. Gupta *et al.* (1995) carried out various experiments to find the most suitable water harvesting techniques for plants/saplings in (a) pits of normal size, (b) pits surrounded by a saucer-shaped depression, (c) pits surrounded by a ring-shaped depression, (d) pits with a sloping side and with a mound in the middle, (e) a trench and mound, (f) deep ploughing. The effect of spacing on sprout growth was studied by Raizada and Padmaiah (1995), after coppicing. The best results were obtained by 3×1 m spacing.

Other Studies

For an agroforestry system, particularly in dry areas, the rate of water consumption and biomass production are important. Chaturvedi *et al.* (1985) gave an account of the farming of neem as a source of fuel wood. Chandrasekharaiah and Prabhakar (1987) studied the harvestable biomass of 4- and 5-year-old trees. It was 15.77 kg for neem, as compared to 70.1 kg for *Dalbergia sisoo*. There was more than 50 percent increase in dry matter production after 8 weeks when 2–3-month-old seedlings were kept with extra carbon dioxide maintained at the concentration of 8000 ppm, twice as much compared to the control (Suraminath *et al.*, 1988). Lysimeter measurements for this were taken by Chaturvedi *et al.* (1988). Mulching using coir pith (a waste material from coconut processing) and fertilizer treatment improved N, P and K concentration in plants and significantly improved soil fertility and moisture content (Balwinder Singh *et al.*, 1988). The effect of mulching under different irrigation levels on the height of tree seedlings was studied by Singh *et al.* (1989), while Gupta (1991) gave an account of the effect of fertilizer application on initial development. Singh *et al.* (1991) studied the response of leaf residues and irrigation on plants. Application of leaf mulch considerably improved growth, even at lower levels of irrigation treatment.

Biomass equation above ground, biomass in dry matter and nutrient content were studied by Bunlyavejchewin (1989), Bunlyavejchewin and Kiratiprayoon (1989) and Bunlyavejchewin *et al.* (1989). Pine needles, water management of transplanted seedlings in arid areas was studied by Burman *et al.* (1991). An average application of 46 liters, full-field capacity per nine-month-old plant at two-week intervals, led to maximum growth and biomass production with no mortality. Brenner *et al.* (1991) determined the daily transpiration rate for the tree in an unstressed wind-break. Marti *et al.* (1991) studied growth and biomass in the coastal saline wasteland of east India. Gupta (1994) described the effect of rain water on biomass production in neem in the Indian desert. In Nigeria, waste water was found to result in enhanced height growth with no apparent harm (Lacuali *et al.*, 1995).

Arya *et al.* (1992) described a sweeping association of *Azadirachta indica* with *Prosopis cineraria*.

Jattan *et al.* (1995, 1995a) suggested various steps for the management of neem plantations; this included pruning the trees for floral initiation and to stop irregular bearing. When fruit shedding is expected, it can be prevented by proper irrigation and the supply of nutrients, by removing shade and shoot tips. Weeding and hoeing, which loosens the soil and helps aeration, have also been found useful.

Various diseases in neem can be controlled, as given by Tiwari (1992).

PLANT DISEASES

Various constituents from neem have been found effective against insect, fungal and even viral pathogens but neem itself is attacked by some of these. The attack may be on seed, on the young seedling, on the leaf, on the root and even on the whole tree.

Wealth of India (1948) mentioned the various fungal diseases of neem caused by *Cercospora leucosticta*, *C. subsessilis*, *Xylaria azadirachtae*, *Fomes* and *Polyporus*. Schmutterer (1990) has given an account of observations on 20 species of insects and 12 species of arthropods pests, along with their biology and geographical distribution. Tewari (1992) and Schmutterer (1995) have given detailed accounts of all these diseases.

The important ones, organ wise, are the following.

Whole Tree

Bacterial wilt by *Pseudomonas solanacaerum* has been described by Diatloff *et al.* (1993). It caused black to brown discoloration in the root and stem with collar rot developing in the advanced stage. Benge (1993) has given an account of the disease called "neem disorder" or "neem decline". It has been seen in an area from Mali and the west to Chad and Cameroon in Africa. The disease is characterized by gradual loss of foliage, a general debilitation of the tree and sometimes death. This is probably caused by the *Verticillium* species. The other disease is pink disease caused by *Corticium salmonicolor*. It is characterized by the formation of a pink incrustation and canker on the stem. Phomopsis twig blight is caused by *Phomopsis*, which attacks the twig (Tewari, 1992).

Seed

Oryzaephilus surinamensis was found infesting the seed by Zongo (1990). The infested seed were entirely hollowed out. Uniyal and Uniyal (1996) collected neem seed from 24 different localities and found them heavily inhabited by seven fungal genera, *Fusarium*, *Aspergillus*, *Penicillium*, *Cephalosporium* spp., *Alternaria* spp., *Pythium* spp. and *Mycelia sterelia*. The dominating species were *Aspergillus niger*, *A. fumigatus* and *Fusarium*.

Seedlings

In a forest nursery, Raghunathan *et al.* (1982) saw thousands of fire ants (*Solenopsis* spp.) attacking the saplings, and completely defoliating them. These ants did not touch the other plants nearby. Sankaran *et al.* (1988) observed leaf spot disease caused by *Colletotrichum capsici* and *Cercospora subsessilis* in 3-month-old seedlings. Leaf blight and stem rot by *Sclerotium rolfsii*, and Web blight by *Rhizoctonia solani*, have also been observed. Twig blight has been reported by Kaushik *et al.* (1993). It caused minute, circular, black to brown lesions on the neck region of 2–3-month-old seedlings. The infection caused the death of seedlings. The causative organism was the *Colletotrichum* state of *Glomerella cingulata*.

Root

Phytophthora cinnamoni has been found attacking the tree worldwide (Benge, 1993). Black to brown discoloration occurred in the root due to *Ganoderma applanatum* (Chakraborty and Kongers, 1995). Ganoderma root rot is caused by *G. lucidum*. Drying of leaves takes place after the root system is damaged (Tewari, 1992).

Leaves

The various diseases reported are leaf spot by *Xanthomonas azadirachti* (Moniz and Raj, 1967), powdery mildew by *Oidium azadirachtae* (Naryanasmy and Ramkrishnan, 1971) and bacterial leaf spot (Nayudu, 1972). Singh and Chohan (1984) observed that *Phoma joylana* caused severe twig canker and shot holes in the leaves. Sankaran *et al.* (1988) for the first time recorded the *Colletotrichum gloeosporioides* state of *Glomerella cingulata*. Jamaludin *et al.* (1988) described Cercospora blight disease, caused by *C. subsessilis*, in winter. It started on the leaves, and spread to the other parts, leading to shedding of leaves and fruit. Seeds so produced were not viable.

Harsh *et al.* (1989) also observed neem spot caused by *Cercospora*. Pillai and Gopi (1990) mentioned that the drying up of the distal parts of the shoot and foliage in the months December to April has become a common phenomenon in south India, which is due to the tea mosquito bug. In addition to this, 16 insect species and one mite, *Calipitrimerus azadirachtae*, were identified. Mehrotra (1990) gave an account of foliar disease, leaf web blight, caused by *Rhizoctonia solani* in forest nurseries. The symptoms of leaf web blight disease are the development of greyish brown blotches, which grow with age and cover the whole leaflet. The infected leaflet remain joined by fungal hyphae, as in the case of a spider's web. The disease appears during the high humidity months. The pathogen is borne in the soil, and climbs up on the leaves, infecting them one by one. Hiremath *et al.* (1991) recorded infection by *Cercospora leucosticta* and *Glomerella cingulata*. Colletotrichum leaf spotting blight and Alternaria leaf spotting blight have been described by Tewari (1992).

Karthikeyan *et al.* (1993) observed neem to be infested by 13 pests; the tea mosquito bug *Hellopeltis antonii*, the mealy bug *Pseudococcus gilbertensis*, the scale insect *Parlatoria orientalis* and the leaf webber *Loboschiza koenigiana* are the important pests. *H. antonii* damaged the entire foliage from October to March.

Alam (1993) noticed that *Curvularia* spp. and *Exserohilum* spp. caused leaf spot patches, resulting in total rot of the leaf. Another dermataceous hyphomycetes was also recorded causing leaf spot.

REVIEWS

Shankar (1988) gave an account of silvipasture studies on neem in India, for forage and fuel purposes, while Siddique (1989) described neem as a multi-purpose species in Pakistan. Social forestry and rural economy have been discussed by Sharma (1989) for fuel, forage and the development of small-scale/cottage industry. The same approach has been given by Adegbehin *et al.* (1990) in Nigeria. Kamo (1990) and Lauridsen *et al.* (1991) studied it as a fast-growing species in Thailand. Hegde (1991) studied the silvopastoral system with a brief account of land utilization. Tewari (1992), in a monograph on neem, devoted a chapter to silviculture and management. Sidhu (1995) gave an account of neem in agroforestry, by combining the woody perennial with annual crops and animals on the same unit of land management to obtain maximum productivity.

REFERENCES

Adegbehin, J.O., Igboanugo, B.I. and Omijeh, J.E. (1990) Potential of agroforestry for sustainable food and wood production in the Savanna area of Nigeria. *Savanna*, **11**, 12–26.

Ahmed, A.A. and Puzari, N.N. (1991) Initial growth and survival of different forest tree species in Assam. *Indian Forester*, **117**, 549–552.

Ahmed, R., Khan, D. and Ismail, S. (1985) Growth of *Azadirachta indica* and *Melia azdirach* at coastal sand using highly saline water for irrigation. *Pakistan Journal of Botany*, **17**, 229–233.

Ahmed, S., Salem, Banofleh and Ma'tong-Munshi (1989) Cultivation of neem (*Azadirachta indica*, Meliaceae) in Saudi Arabia. *Economic Botany*, **43**, 35–38.

Alam, M.S. (1993) Some fungal diseases in the leaf of *Azadirachta indica*. (abstr.) *Proc. World Neem Conference*, 24–28 February, 1993, Bangalore, India.

Allan, E.J., Eeswara, J.P., Johnson, S., Mordue, A.J., Morgan, E.D. and Stuchbury, T. (1994) The production of azadirachtin by *in vitro* tissue culture of neem *Azadirachta indica*. *Pesticide Science*, **42**, 147–152.

Arya, S., Toky, O.P., Harris, P.J.C. and Harris, S.M. (1992) Prosopis cineraria and Azadirachta indica a sweeping association. *Agroforestry Today*, **4**, 9–10.

Badi, K.H., Ahmed, A.E.H. and Bayoumi, A.A.M.S. (1989) *The Forest of Sudan*. Ministry of Agriculture Khartoum, Sudan.

Balvinder Singh, Gupta, G.N. and Prasad, K.G. (1988) Use of mulches in establishment and growth of tree species on dry lands. *Indian Forestry*, **114**, 307–316.

Benge, M. (1993) Neem Decline. *Neem News*, **1**, 5–6.

Berjak, P., Campbell, G.K., Farrant, J.M., Omondi-Oloo, W. and Pammenter, N.W. (1995) Response of seeds of *Azadirachta indica* (neem) to short term storage under ambient or chilled condition. *Seed Science and Technology*, **23**, 779–392.

Bhardwaj, S.D. and Gurdev Chand (1995) Storage of neem seeds. Potential and limitations for germplasm conservation. *Indian Forester*, **121**, 1007–1011.

Bist, R.P. and Toky, O.P. (1993) Growth pattern and architectural analysis of nine multipurpose trees in arid region of India. *Canadian Journal of Forest Research*, **23**, 722–730.

Bose, R.K. and Bandoyopadhyay, S.K. (1986) Economic of energy plantation in alkali soils of Indian semi-arid regions. *Biomass*, **11**, 51–60.

Brenner, A.J., Jarvis, P.G. and Vandenbeldt, R.J. (1991) *Transpiration from a neem network in the Sahel*. IAHS Publications No. 199, 375–385.

Brenner, A.J., Jarvis, P.G. and Beldt, R.J. van-den (1995) Wind-break crop interaction in the Sahel 1. Dependence of shelter on field conditions. *Agricultural and Forest Meteorology*, **75**, 215–239.

Branney, P. (1989) *Propagation of tree species for aforestation in northern Sudan*. Northern Region Irrigation Project Forestry Development, London.

Bunlyavejchewin, S. (1989) Primary production of plots of five young closed spaced fast growing tree species. III Dry matter and nutrient content of litter fall. *Natural History Bulletin of the Siam Society*, **37**, 65–73.

Bunlyavejchewin, S. and Kiratiprayoon, S. (1989) Primary production of plots of five young closed spaced fast growing tree species. I Biomass equations. *Natural History Bulletin of the Siam Society*, **37**, 47–56.

Bunlyavejchewin, S., Kiratiprayoon, S. and Kumun, T. (1989) Primary production of plots of five young closed spaced fast growing tree species. II Above ground biomass, nutrient and energy content. *Natural History Bulletin of the Siam Society*, **37**, 57–63.

Burman, U., Kathuju, S., Garg, B.K. and Lahiri, A.N. (1991) Water management of transplanted seedlings of Azadirachta indica in arid areas. *Forest Ecology and Management*, **40**, 51–63.

Campolucci, P. and Paolim, C. (1990) Desertification control in the Sahel region in low cost large scale afforestation techniques (in Italian). *Note Ternidie Centro Sperimentazione Agricolaa e Forestale Instituto di Sperimentazione per la Proppicolture* No. 24, 1–2.

Chaisurisri, K., Ponay, B. and Wasuwanich, P. (1986) Storage of *Azadirachta indica* A. Juss, seed. *Embryon*, **21**, 19–27.

Chakraborty, R. and Konger, G. (1995) Root rot of neem. *Indian Forester*, **121**, 1081–1083.

Champion, H.C. and Seth, S.K. (1968) *A Survey of Forest Types of India*. The Manager of Publications, Government Press, Delhi, India.

Chander, H., Singh, R.S. and Mandal, B.S. (1996) Effect of auxin on number of leaves per cuttings and length of main branch in neem *Azadirachta indica*. *Annals of Biology*, **12**, 57–61.

Chandrasekharaiah, A.M. and Prabhakar, A.S. (1987) Growth and its analysis of tree species in agroforestry system. *Journal of Farming System*, **3**, 17–21.

Chaney, W.R. and Knudson, D.M. (1988) Germination of seeds of *Azadirachta indica* enhanced by endocarp removal. *International Tree Crop Journal*, **5**, 153–161.

Chaturvedi, A.N. (1985) *Fuelwood farming on degraded lands in the Gangetic plains*. No. 50, 1–52, U.P. Forest Bulletin, Lucknow, India.

Chaturvedi, A.N., Bhatt, D.N., Singh, U.N. and Gupta, N.N. (1985) Response of certain tree species to various pH levels under pot culture. *Van Vigyan*, **23**, 79–84.

Chaturvedi, A.N., Sharma, S.C. and Srivastava, R. (1988) Water consumption and biomass production of some forest tree species. *International Tree Crops Journal*, **5**, 71–76.

Chaturvedi, M.D. (1955) Nim, hardy, thrifty and useful. *Indian Farming*, **9**, 32–33.

Chinnamani, S. (1993) Neem research at ICAR. Genetic improvement of neem. Strategies for the future. *Proceedings of International Consultation on Neem Improvement*, Bangkok, Thailand, pp. 11–17.

Chiu, S.F. (1993) Investigations on botanical insecticides in South China—an update. *Botanical Pesticides in Integrated Pest Management*. Indian Society of Tobacco Science, Rajahmundry, India, pp. 134–137.

Chourasia, H.K. and Roy, A.K. (1991) Effect of temperature, relative humidity and light on aflatoxin production in neem and Datura seeds. *International Journal of Pharmacology*, **29**, 197–202.

Chowdhury, M.K. (1992) Kendobona Development Project—a novel approach to wasteland reclamation. *Indian Forester*, **118**, 879–886.

Dalal, M.R., Dahiyo, D.S., Sarmah, M.K. and Narwal, S.S. (1992) Suppression efforts of arid zone trees on plant stand and growth of crops. *Proceedings of First National Symposium on Allelopathy in Agro Ecosystems* (Agriculture and Forestry) Feb. 12, 1992, Hissar, India.

Desh Raj (1990) Experience in wasteland development: a case study. *Proceedings of the National Solar Energy Convention*, 1–3 Dec., 1989, Udaipur, India.

Diatloff, A., Wood, B.A. and Wright, D.G. (1993) Bacterial wilt of neem tree caused by Pseudomonas solanacerum. *Australian Plant Pathology*, **22**, 1.

Drechsel, P., Glaser, B. and Zech, W. (1991) Effect of four multipurpose tree species on soil amelioration during tree fallow in central Togo. *Agroforestry System*, **16**, 193–202.

Drew, R.A. (1993) Clonal propagation of neem by tissue culture.(abstr.) *Proc. World Neem Conference*, Banglore, India 24–28 Feb., 1993.

Ekundayo, O. (1983) Biosynthesis of nimbolide in *Azadirachta indica* A. Juss. from (2-14C) mevalonate and (2-14C) acetate. *Zeitscheift fur Pflanzenphysiologie*, **112**, 139–146.

Gautam, V.K., Nanda, K. and Gupta, S.C. (1993) Development of shoots and roots in anther derived callus of Azadirachta indica A.Juss—a medicinal tree. *Plant Cell-Tissue and Organ Culture*, **34**, 13–18.

Gill, H.S. and Abrol, I.P. (1991) Salt affected soils, their afforestation and ameliorating influence. *International Tree Crops Journal*, **30**, 239–260.

Gupta, G.N. (1991) Effect of mulching and fertilizer application on initial development of some tree species. *Forest Ecology and Management*, **44**, 211–221.

Gupta, G.N. (1992) Growth of nursery plants of *Azadirachta indica* as influenced by soil mixture and fertilizers in arid region. *Van Vigyan*, **30**, 59–63.

Gupta, G.N. (1994) Influence of rain water harvesting and conservation practices on growth and biomass production of Azadirachta indica in the Indian desert. *Forest Ecology and Management*, **70**, 329–339.

Gupta, G.N. (1995) Rain water management for tree planting in the Indian desert. *Journal of Arid Environment*, **31**, 219–235.

Gupta, G.N., Limba, N.K. and Gupta, P.K. (1995) Micro-catchment water harvesting for raising neem in arid regions. *Indian Forester*, **121**, 1022–1032.

Gupta, G.N., Mohan, S. and Prasad, K.G. (1987) Salt tolerance of selected tree seedlings. *Journal of Tropical Forestry*, **3**, 217–227.

Harikrishnan, M. (1993) Coastal agroforestry and its sustainable utilisation. In M.S. Swaminathan and R. Ramesh (eds.) *Workshop on Sustainable Management of Coastal Ecosystem*, 1991, 71–83.

Harsh, L.N., Tewari, J.C., Bueman, U. and Sharma, S.K. (1992) Agroforestry for arid regions. *Indian Farming*, **42**, 32–37.

Harsh, N.S.K., Tiwari, C.K. and Nath, V. (1989) Foliage disease in forest nurseries and their control. *Journal Tropical Forestry*, **5**, 66–69.

Hedge, N.G. (1991) On farm afforestation research in extension programme: methods used by BAIF in India. In S.J. Schers (ed.) Methods for participating on farm agroforestry research. *Agroforestry Systems*, **15**, 203–216.

Hiremath, S.V., Kulkarni, S., Hegde, R.K., Pall, B.S. and Bhat, M.N. (1991) New host records of fungi. *Current Research University of Agricultural Sciences*, Bangalore, India, **20**, 93–94.

Jha, M. and Choudhuri, L.D. (1990) Trial on stump planting of some tree species. *Indian Forester*, **116**, 283–285.

Jha, M. and Chaudhuri, L.D. (1995) A note on increasing the viability of neem seed. *Indian Forester*, **121**, 1085–1086.

Jamaludin, Dadwal, V.S. and Sima Share (1988) Cercospora blight disease of Melia Azadirachta. *Indian Forester*, **114**, 894–895.

Jattan, S.S., Shashi Kumar, Pujar, G. and Bisht, N.S. (1995) Prespective in intensive management of neem plantations. *Indian Forester*, **121**, 981–988.

Jattan, S.S., Shasi Kumar, Pujar, G. and Kumar, S. (1995a) Margoculture a vital component for plant health in future. *Indian Forester*, **121**, 993–995.

Joarder, N., Joarder, O.I., Islam, R., Zaman, A. and Biswas, B.K. (1993) *In vitro* response of nucellar tissue and plant regeneration from somatic embryogenesis from cultured cotyledons of neem.(abstr.) *Proc. World Neem Conference*, Bangalore, India. 24–28 February, 1993.

Joshi, M.S. and Thengane, S.R. (1996) *In vitro* propagation of *Azadirachta indica* A. Juss.(neem) by shoot proliferation. *Indian Journal of Experimental Biology*, **34**, 480–482.

Joshi, P.C. and Prakash, O. (1992) Allelopathic effects of litter and extract of some tree species on germination and seedling growth of agricultural crop. In T.P. Narwal (ed.) *Proc. First National Symp. Allelopathy in Agroeco-systems, Agriculture and Forestry*, Feb. 12–14, 1992, Haryana Agricultural University, Hissar, India.

Kalla, J.C., Gyan, C., Vyas, D.L. and Gehlot, N.S. (1978) Techno-economic felling cycles for selected energy plantation species in the arid areas of Western Rajasthan. *Annals of Arid Zone*, **18**, 42–57.

Kamo, K. (1990) Productivity of fast growing tree in Thailand. *Tropical Forestry*, **19**, 26–34.

Karthikeyan, K., Rangarajan, A.V. and Velusamy, R. (1993) Major pests of neem and their management in southern Tamilnadu. (abstr.) *Proc. World Neem Conference*, Bangalore, India 24–28 February, 1993.

Kaushik, J.C., Bisla, S.S. and Singh, D.P. (1993) A new disease of neem seedling. (abstr.) *Proc. World Neem Conference*, Bangalore, India 24–28 February, 1993.

Ketkar, C.M. (1982) Properties and uses of neem its products and by products. In C.K. Atal and B.M. Kapur (eds.) *Cultivation and Utilization of Medicinal Plants*. Regional Research Laboratory, Jammu, India.

Kinhal, G.A. (1968) Early planting and critical watering—an effective method of wasteland afforestation. *Advance in Forestry Research in India*, **1**, 125–134.

Kismul, H. (1989) *What can be learned from the Fellata people about tree planting*. Field note Asian Productivity Council, Minato-Ku, Japan.

Kumaran, K., Surendran, C. and Rai, R.S.V. (1993) Variation studies and heritable components of seed parameters in neem Azadirachta indica A. Juss. (abstr.) *Proc. World Neem Conf.*, 24–28 February, 1993, Bangalore, India.

Lacuali, G., Delisle, C.E. and Vincent, G. (1995) Preliminary study of the reuse of waste water for irrigation in forest nurseries in Niger. *Secheresse*, **6**, 210–205.

Laskar, S. and Datta, M. (1992) The cover improves soil fertility and water retention. *Indian Farming*, **41**, 10–17.

Lauridsen, E.B., Kanchanaburangura, C. and Boonsermsuk, S. (1991) Neem (*Azadirachta indica* A. Juss) in Thailand. *Forest Genetic Resource Information*, **19**, 25–33.

Lewis, W.H. and Elvin-Lewis, M.P.H. (1983) Neem (*Azadirachta indica*) cultivated in Haiti. *Economic Botany*, **37**, 69–70.

Maithani, G.P., Bahuguna, V.K. and Pyare Lal (1988) Studies on nursery techniques (method of sowing and optimum depth of sowing) of *Azadirachta indica* A. Juss under north Indian moist tropical conditions. *Indian Forester*, **114**, 440–445.

Maithani, G.P., Bahuguna, V.K., Rawat, M.M.S. and Sood, O.P. (1989) Fruit maturity and interrelated effects of temperature and container on longevity of neem, *Azadirachta indica* seed. *Indian Forester*, **115**, 89–92.

Marti, S., Mandal, N.N., Mandal, S. and Chaudhuri, P.P. (1991) Growth and biomass productivity of trees planted at different spacing in the coastal saline wasteland of Sunderbans. *Indian Journal of Landscape System and Ecological Studies*, **14**, 114–120.

Martinez, H.A. (1987) Silviculture of various species of multi-purpose tree III (in Spanish). *Chasqui Costa Rica*, **14**, 11–17.

Matig, O.E. (1989) Attempts to rehabilitate the stable soils forest plantation in north Cameroon. Trees for development in sub-Saharan Africa. *Proc. Regional Seminar held International Foundation of Science*, Kenya, Feb. 20–25, 1989, 213–227.

Meena, J.N., Gupta, J.N. and Ram, T. (1995) Influence of soil technique on early growth of trees in arid zone. *Annals of Forestry*, **3**, 120–128.

Mehrotra, M.D. (1990) Rhizoctonia solanii, a potential dangerous pathogen of Khasi pine and hardwoods in forest nurseries in India. *European Journal of Plant Pathology*, **20**, 329–330.

Mehta, K.K. (1989) Grow trees on wasteland alkali soils. *Indian Farming*, **39**, 27–28.

Mohan, S., Prasad, K.G. and Gupta, G.N. (1990) Fertilizer response of selected social forestry species under varying soil texture. *Indian Forester*, **116**, 49–57.

Mohinderpal (1995) Clonal approach for yield improvement in neem. Strategies and protocols for selective use of genetic diversity. *Indian Forester*, **121**, 1033–1038.

Moniz, K. and Raj, H.H. (1967) Xanthomonas azadirachti causing leaf spot in Azadirachta indica A. Juss. *Indian Journal of Microbiology*, **7**, 159–160.

Muthana, K.D., Mohinder Singh, Mertia, R.S. and Arora, G.D. (1984) Shelterbelt plantations in arid regions. *Indian Farming*, **33**, 19–2.

Nagaraju Kumar, Y.S.A., Ramachandran, K.S., Syed Ibrahim, Ranganna, B. and Krishnappa, A.M. (1992) Contour 'V' ditch—a new technique in forest plantation. *Current Research—University of Agricultural Sciences*, Bangalore, **21**, 148–150.

Nagaveni, H.C., Ananthapadmanabha, H.S. and Rai, S.N. (1987) Note on the extension of maturity of *Azadirachta indica*. *Myforest*, **23**, 245.

Naina, N.S., Gupta, P.K. and Mascarenhas, A.F. (1989) Genetic transformation an regulation of transgenic neem, *Azadirachta indica*, plants using Agrobacterium tumefaciens. *Current Science*, **58**, 184–187.

Naryanasmy, P. and Ramkrishnan, K. (1971) Powdery mildew of Coimbatore. *Journal of the Madras University*, **38**, 84–99.

Nayudu, M.V. (1972) Pseudomonas viticola sp.nov. Incident of a new bacterial disease of grape vine. *Phytopathology Journal*, **73**, 183–186.

Nehra, O.P., Oswal, M.C. and Faroda, A.S. (1987) Management of fodder trees in Haryana. *Indian Farming*, **37**, 31–33.

Newton, R.J., Jain, S.M. and Gupta, P.K. (eds.) (1995) *Somatic embryogenesis in woody plants*. Vol. II *Angiosperms*. Klumer Academic Publishers, Netharlands.

Nirmala Kumari, A., Ramaswamy, N.M. and Rangaswamy, S.R. (1993) Tissue culture studies and neem *Azadirachta indica*.(abstr.) *Proc. World Neem Conference*, 24–28 February, 1993, Bangalore, India.

Oboho, E.G. Nwoboshi, L.C. (1991) Windbreaks: how well do they really work. *Agroforestry Today*, **3**, 15–16.

Oguntala, A.B. (1989) The climatic aspects of 1982-83 wildfire in Nigeria. Meteorology and Agroforestry. *Proc. Int. Work. on Application of Metereology to Agroforestry Systems, Planning and Management*, Nairobi, Kenya, 9–13 Feb., 1987, pp. 539–546.

Onyewotu, L.O.Z. (1985) Establishment consideration for optimum spacing of shelter belts in the Sahel and Sudan zones of Nigeria. In J.A. Okojie and O.O. Okoro (eds.) *Proc. 15th Annual Conf. Forestry Assoc., Nigeria. Forest Association of Nigeria*. Lagos.

Pal, M., Adarsh Kumar, Bakshi, M. and Bhandari, H.C.S. (1994) Cheap non-auxin chemicals for rooting nodal segments of neem (*Azadirachta indica*) *Indian Forester*, **12**, 138–141.

Pal, M., Badole, K.C. and Bhandari, H.C.S. (1992) Stimulation of adventitious root regeneration on leafy cuttings of neem (*Azadirachta indica*) by auxins and phenols. *Indian Journal of Forestry*, **15**, 68–70.

Panchan, S., Katawatin, R. and Srisataporn, P. (1989) Effect of salinity on growth of fast growing trees. *Kaen Kast Khen Kaen Agricultural Journal*, **17**, 91–99.

Pillai, S.R.M. and Gopi, K.C. (1990) Seasonal drying up of distal shoot of neem Azadirachta indica A. Juss and important insect pests associated with it. *Myforest*, **26**, 33–50.

Pong-Anant, Wongamanee, K.C., Wasuwanich, P. and Kijkar, S. (1989) *Grafting of Azadirachta indica A. Juss*. Muak-Lek, Saraburi Thailand Asean Canadian Forest Tree Seed Centre, Thailand.

Ponnuswamy, A.S. and Karivaratharaju, T.V. (1993) Standardisation of seed grading as upgrading techniques for neem seed.(abstr,) *Proc. World Neem Conference*, 24–28 February, 1993, Bangalore, India.

Ponnuswamy, A.S. and Karivaratharaju, T.V. (1993a) Seed pelleting. (abstr.) *Proc. World Neem Conference*, 24–28 February, 1993, Bangalore, India.

Ponnuswamy, A.S., Karivartharaju, T.V., Balagurunathan, R. and Parameswaran, S. (1993) Detection of inhibitors in the fruits of neem. (abstr.) *Proc. World Neem Conference*, 24–28 February, 1993, Bangalore, India.

Ponnuswamy, A.S., Rai, R.S.V., Surendran, C. and Karivaratharaju, T.V. (1991) Studies on mounting seed longevity and the effect of seed grade in neem *Azadirachta indica*. *Journal Tropical Forest Science*, **3**, 285–290.

Prasad, J. (1941) Germination of nim seed (Azadirachta indica A. Juss). *Indian Forester*, 67.

Prasad, R. and Dhuria, S.S. (1989) Reclamation of iron ore mined out areas: biomass production efficiency of species. *Journal Tropical Forestry*, **5**, 51–56.

Pratap, N. and Jaiswal, V.S. (1985) Plantlet regeneration from leaflet callus of Azadirachta indica A. Juss. *Journal of Tree Sciences*, **4**, 65–68.

Preetha, N., Yasodha, R. and Gurumurthi, K. (1995) Peroxidase isoenzyme as markers of organogenesis in Azadirachta indica A. Juss. *Indian Journal of Plant Physiology*, **38**, 92–93.

Pukittayacamcee, P., Biontawee, B., Wasuwanich, P. and Boonarutee, P. (1995) Effect of fruit maturity, depulping techniques and drying conditions on germination of Azadirachta indica var. siamensis seed. *ASEAN Forest Tree Seed Center Project* No. **33**, 111–151.

Puri, H.S. and Hardman, R. (1980) Effect of various drying conditions on the outer seed surface of fenugreek. *Proc. All India Symposium on Current Researches on Plant Sciences*, 22–24 April, 1980, Panjab University, Chandigarh, India.

Puri, S., Bangarwa, K.S. and Singh, S. (1995) Influence of multi-purpose trees on agricultural crops in arid regions of Haryana, India. *Journal of Arid Environments*, **30**, 441–451.

Raddi, A.G. (1981) Integration of forage forestry in afforestation programmes of Maharashtra State. *Indian Journal of Range Management*, **2**, 81–85.

Radhamani, J., Chaudhuri, R. and Chander, K.P.S. (1990) Inhibition of seed germination by the endocarp in neem. *Indian Journal of Plant Genetic Resources*, **3**, 35–40.

Radwansky, S.A. and Wickens, G.E. (1981) Vegetative fallows and potential value of neem (*Azadirachta indica*) tree in the tropics. *Economic Botany*, **35**, 398–414.

Raghunathan, T.A.V.S., Alam, M.A. and Venkaiah, K. (1982) Fire ant (Solenopsis spp). *Indian Forester*, **108**, 375.

Raizada, A. and Padamaiah, M. (1995) Coppice growth from tree species growing in energy plantation, effect of spacing. *Indian Forester*, **121**, 613–619.

Rakesh Kumar and Bangarwa, K.S. (1996) Seed storability in *Azadirachta indica* A. Juss. *Annals of Biology, Ludhiana*, **12**, 62–66.

Rakesh Kumar and Bangarwa, K.S. (1996a) Seed collection and pre-sowing handling methods in Azadirachta indica A. Juss. *Annals of Biology, Ludhiana*, **12**, 67–70.

Ramesh Kumar and Padhya, M.A. (1988) Isolation of nimbin from Azadirachta indica leaves and its callus culture. *Indian Drugs*, **25**, 526–527.

Ramesh Kumar and Padhya, M.A. (1990) *In vitro* propagation of neem Azadirachta indica A. Juss. from leaf discs. *Indian Journal of Experimental Biology*, **28**, 932–935.

Ramesh Kumar and Padhya, M.A. (1993) De novo synthesis of secondary metabolites in cultured cells of neem. (abstr.) *Proc. World Neem Conference*, 24–28 February, 1993, Bangalore, India.

Randhawa, N.S. and Parmar, S.S. (eds.) (1993) *Neem Research and development*. Publication No. 3, Society of Pesticide Science, India.

Rao, H.S. (1958) Vegetative propagation and forest tree improvement. *Indian Forester*, **42**, 199–204.

Rao, M.V.S., Rao, Y.V., Rao, Y.S. and Manga, V. (1988) Induction of growth callus in *Azadirachta indica* A. Juss. *Crop Improvement*, **15**, 203–205.

Read, M.D. and French, J.H. (eds.) (1993) Genetic improvement of neem: Strategies for the future. *Proc. Int. Consultation held at Kasetart University*, Bangkok, Thailand, 18–22 Jan., 1993.

Roederer, Y. (1991) Forestry and agroforestry experiments in the dry zone of the west coast of Reunion (in French). *Numero Special La Reunion' Bois et Forest des Tropiques* No. **229**, 51–60.

Sankaran, K.V., Balasundran, M. and Sharma, J.K. (1986) Seedling disease of *Azadirachta indica* in Kerala, India. *European Journal of Forest Pathology*, **165**, 324–328.

Sankaran, K.V., Florence, E.J.M. and Sharma, J.K. (1988) Foliar disease of some forest tree in Kerala – new record. *Indian Journal of Forestry*, **11**, 104–107.

Sanyal, M., Das, A., Banerjee, M. and Datta, P.C. (1981) *In vitro* hormone induced chemical and histological differentiation in stem callus of neem, *Azadirachta indica* A. Juss. *Indian Journal of Experimental Biology*, **19**, 1067–1068.

Sanyal, M., Mukherji, A. and Datta, P.C. (1986) Evaluation of root bark of *Azadirachta indica* (Meliaceae). *Journal of Plant Anatomy and Morphology*, **3**, 13.

Sanyal, M., Mukherji, A. and Datta, P.C. (1988) Effect of glycine on *in vitro*, biosynthesis of nimbin and beta sitosterol in tissues of *Azadirachta indica*. *Current Science, India*, **57**, 40–41.

Sarkar, M.S. and Datta, P.C. (1986) Age factor in biosynthesis of nimbin and sitosterol in the bark and callus of *Azadirachta indica*. *Indian Drugs*, **24**, 62–63.

Sastry, T.C. and Kanathekar, K.V. (1990) *Plants for reclamation of waste land*. Publication and Information Directorate, CSIR, New Delhi, India.

Schulz, F.A. (1984) Tissue culture of *Azadirachta indica*. *Proc. 2nd Inter. Neem Conference*, 25 May 1983, Rauschholzhausen (Germany).

Schmutterer, H. (1990) Observation on pests of *Azadirachta indica* and some species of Melia (in German). *Journal of Applied Entomology*, **109**, 390–400.

Schmutterer, H. (1995) *Neem Tree, source of unique natural products for integrated pest management, medicine, industry and other purposes*. VCH Verlag. Weinhim, Germany.

Shaikh, M.H.A. (1992) Ecological approach to wasteland development. *Myforest*, **28**, 123–128.

Shaikh, M.H.A. (1993) Ecological approach to dry zone afforestation. *Indian Forester*, **119**, 530–535.

Shankar, V. (1988) *Silvipasture research in India*. In Panjab Singh (ed.) *Pasture and forage crop research—A state of knowledge report*. Third International Rangeland Congress, New Delhi, India, pp. 291–304.

Shanmungavelu, K.G. (1967) A note on the air layer of Eugenia (Eugenia jambolana) and Neem (*Azadirachta indica*). *South Indian Horticulture, Coimbatore, India*, **15**, 70–71.

Sharma, S.C. (1989) Social forestry and rural economy. *Journal of Tropical Forestry*, **5**, 12–16.

Sharma, S.D., Prasad, K.G., Singh, R.K. and Malik, N. (1993) Development of soil technology for afforestation of sodic soils in Haryana, India. *Annals of Forestry*, **1**, 168–177.

Siddique, K.M. (1989) Review of research on multipurpose tree species in Pakistan. *Pakistan Journal of Forestry*, **39**, 165–170.

Sidhu, D.S. (1995) Neem in agroforestry as a source of plant derived chemicals for pest management. *Indian Forester*, **121**, 1012–1021.

Sieder, P. (1983) Large scale afforestation in Upper Volta (in German). *Forst und Holzwirt*, **38**, 112, 114, 116–118, 120.

Singh, B.G., Mahadevan, N.P., Shanthi, K., Geetha, S. and Manimuthu, L.(1995) Multiple seedling development in neem. *Azadirachta indica*. *Indian Forester*, **121**, 1049–1052.

Singh, I. and Chohan, J.S. (1984) Phoma joylana—a new pathogen on neem, *Azadirachta indica*. *Indian Forester*, **110**, 1058–1060.

Singh, S.B., Pramod Kumar, Prasad, K.G. and Kumar, P. (1991) Response of various tree species to irrigation regimes and leaf residue management. *Journal Indian Society of Soil Sciences*, **39**, 229–232.

Singh, V.R. (1991) Use of intensive soil working techniques in afforestation of waste land. *Indian Forester*, **117**, 515–520.

Sivaganam, K., Vinaya Rai, R.S., Swaminathan, C. and Sarendran, C. (1989) Studies on rooting response to growth regulators in *Azadirachta indica* A. Juss. *Proceed. Seminar on Vegetative Propagtion*, 27–28 July, 1989, Coimbatore (India).

Sivasamy, M. and Karivaratharyu (1993) Azimuth (Directional) influence on seed quality of neem. (abstr.) *Proc. World Neem Conf.*, Banglore, India 24–28 February 1983.

Smith, J. (1939) Germination of neem seed. *Indian Forester*, **65**, 3.

Spaak, J.D. (1990) Afforestation on Cape Verde—why, how and for whom? (in French) *Bois et Forets des Tropique*, **225**, 47–54.

Stephen, B., Murrugandam and Dyanandan, P. (1993) Neem seedling. growth potential of different organs. (abstr.) *Proc. World Neem Conf.*, 24–28 February 1983, Banglore, India.

Suraminath, M.H., Kushalpa, K.A. and Vaidya, B.A. (1988) Studies on the effect of enrichment of carbon-di-oxide (CO_2) on the growth and dry matter production of different forestry species. *My forest*, **24**, 6–11.

Swaminathan, R.S. and Surendram, C. (1989) Studies on rooting response to growth regulators in *Azadirachta indica* A. Juss. *Proc. Seminar on Vegetative Propagation*, Coimbattore, India, 27–28 July, 1989.

Tampabolon, A.P. and Alrasyid (1989) The neem tree and its developmental prospects in rainfed zones in Indonesia. *Duta Rimba*, **15**, 3–12, 109–110.

Tewari, D.N. (1992) *Monograph on Neem Azadirachta indica A. Juss.* International Book Distributors, Dehradun, India.

Thiagarajan, M. and Muarali, P.M. (1994) Optimum condition for embryo culture of *Azadirachta indica* A. Juss. *Indian Forester*, **120**, 500–503.

Thomossey, J.P. (1991) Natural resources in Chad (in French) *Bois et Forets des Tropiques*, **228**, 49–62.

Toky, O.P. and Bist, R.P. (1992) Observation on the rooting patterns of some agriforestry trees in arid region of north western India. *Agroforestry Systems*, **18**, 245–263.

Ujhmura, E. (1986) Green shelterbelts and afforestation in semi-arid zones—a case study of Nigeria. *Tropical Forestry*, **5**, 42–51.

Umarani, R., Bharathi, A. and Karivaratharaju (1993) Influence of provenance on seed quality in neem (*Azadirachta indica*). (abstr.) *Proc. World Neem Conference*, Banglore, India, 24–28 February, 1993.

Uniyal, K. and Uniyal, D.P. (1996) Mycoflora of neem seed. *Indian Forester*, **122**, 423.

Vanangamudi, K., Padmavathi, S., Manonmani, V. and Surendran, C. (1993) Effect of pelting with biofertilizers, biocides and nutrients on the viability and vigour of neem seed (abstr.) *Proc. World Neem Conference*, Banglore, India, 24–28 February, 1993.

Venkatesh, D.A., Dundiaiah, M. and Mahadevanmurthy, S. (1990) Viability of neem seed. *My forest*, **26**, 369–370.

Verinumbe, I. (1991) Agroforestry development in north eastern Nigeria. In P.G. Jaevis (ed.) *Agroforestry: Principles and Practices. Proc. Int. Conf.*, *Edinburgh*, UK, 23–28 July, 1989.

Verma, R.C., Dillon, R.S. and Singh, V.P. (1996) Effect of auxins on rooting of neem cuttings in spring season. *Annals of Botany*, **12**, 53–56.

Vijayalakshmi, K., Radha, K.S. and Vabdana, Shiva (1995) *Neem—a User's Manual.* Research Foundation for Science, Technology and Natural Resource Policy, New Delhi, India.

Werner, D. and Muller, P. (1990) Fast growing trees and nitrogen fixing trees. *Int. Conf. Marburg*, 8–12 Oct., 1989, Gustav Fischer Verlag, Stuttgart, Germany.

Wealth of India (1948), Publication Division, CSIR, New Delhi.

Zongo, J.O. (1990) Insect pests of nuts and leaves of neem. *Sahel PV Info*, **27**, 19–20.

5. PLANT RAW MATERIAL

The seed as a source of azadirachtin and the oil are the main raw material of the industry. To some extent, bark is used in Ayurvedic and Homoeopathic medicines and leaves as a herbal remedy. An account of these is given separately.

THE SEED

Collection

During field studies, it was observed that the neem tree has a sporadic distribution; there are no neem forests, consequently only a small quantity of seed could be collected from a particular area. Mostly seed is collected by picking ripe seed which falls on the ground by itself but sometimes, when the fruit turns yellow, the branches are shaken or beaten by the collector climbing the tree. By this act the fruit falls on the ground and may be collected on a sheet of cloth or made into a heap by sweeping the ground with a straw broom. The neem fruit matures in most places during the rainy months of monsoon, when there is not only high atmospheric humidity and water logging of the soil but water everywhere on the ground accompanied by flood and rivers in spate. This makes the job of collecting, drying and transporting good quality neem seed very difficult. In the absence of any known post-harvest technology, and in the above conditions, in earlier days most of the seed deteriorated during and after collection. A good infrastructure is required for the transportation of seed from rural areas to the markets or to collection centres in the towns.

Harvesting Methods

Mitra (1963) carried out a detailed literature survey and gave proper post-harvest technology for the seed, so that good quality raw material is obtained. It was seen that in the markets seed was brought in the following stages: (1) dry, (2) semi-dry, (3) dry but wet, (4) damaged and (5) fresh.

Drying Methods

For proper drying of the seed, Mitra (1963) suggested that they should be regularly moved for aeration. This can be done in the field by making four quadrants. The seed should be stored on racks of bamboo or on benches for movement of air. Every day one heap should be moved from one quadrant to another. When the seed are at the optimum level of moisture and there is no danger of fungal contamination after storage, they should be de-pulped.

The process of de-pulping involves soaking the seed for 4 to 5 days in water. When the pulp becomes soft, the seed should be fed into a revolving drum which has a central shaft with pedals. The inner side of the drum has baffles. After feeding the seed, the drum can be rotated by hand or by other mechanical means so that the outer soft seed coat gets separated from the inner hard seed, leaving some pulp only.

The above treatment is for large-scale operations but for smaller quantities, the seed are de-pulped by macerating them by hand after the addition of wood ash. By this treatment the mucilaginous pulp loses its mucilaginous nature, and the removal of the seed coat becomes easier. In villages in south India, the seed coats from fresh seed are removed by collecting them in a heap on the dry ground and placing old jute bags on it. The seed are macerated by foot, by standing on the heap. By foot movement, pressure is exerted on the seed pulp which causes the separation of the seed coat. The mucilaginous seed pulp sticks to the jute bag.

For proper storage, it is essential that seed are free of pulp. The pulp not only absorbs moisture during storage, but acts as a medium for fungal growth, because of its nutrient nature. Seed can be cleaned of pulp by repeatedly washing them with water. In a method suggested by Mitra (1963), an inclined metallic sieve is used. The de-pulped seed are spread on it with a shovel and a jet of water is sprinkled on them by force. This separates the remnants of pulp on the seed, which escapes through the sieve. The clean seed settles down.

The seed are dried once more in dry air or in dryers.

Storage Conditions

As given in Chapter 4 on cultivation, quite a number of studies have been conducted on the proper storage of seed so that there is no deterioration. It has been observed that seed made free from pulp by repeatedly washing them with water have less chance of getting contaminated. Both high humidity and high temperature are conducive to fungal growth within and outside the seed. If the seed are stored in a dry atmosphere and with good aeration, the deterioration of seed can be prevented to a major extent.

Morphology of the Fruit and Seed

Fruit: It is an ovoid drupe, yellow to brown when ripe, epicarp thin, mesocarp with scanty mucilaginous sweetish pulp, endocarp hard enclosing the seed. The fruit gets darker in color and wrinkled on drying (Fig. 9A). When the seed coat is removed, seed with hard endocarp get separated (Fig. 9b1; b2). Seed exalbuminous, plano-convex, notched at the base (Fig. 9c1), very often the seed kernel is covered over by brownish fungal spores (Fig. 9c2) or is totally replaced by these spores. The seed length varied between 11 and 18 mm, with width 4.5–8.5 mm and weight 100–530 mg.

THE BARK

Collection

It is collected mainly from the felled trees by stripping it off from the stem and dried in shade. It may also be obtained from the standing tree, but in that case the precaution is taken that only a small portion of the stem is exposed, otherwise the tree may die of

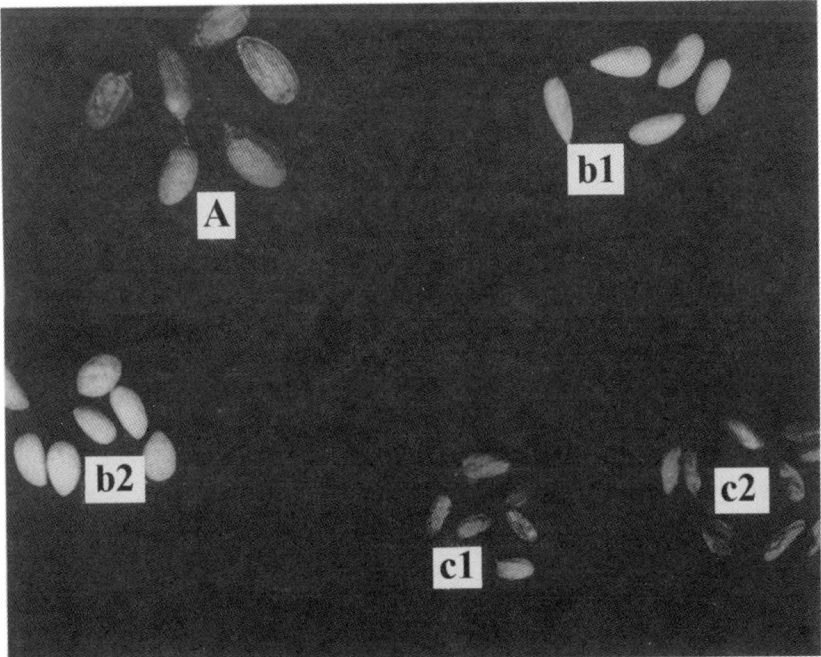

Figure 9 *Azadirachta indica*. A, whole dired seed, showing shriveled outer surface; B. seed with outer coat and pulp removed; (b1) pear shaped, (b2) globular and ovoid; C. seed kernel; (c1) fungal infested, (c2) normal

dehydration. Where it is essential to remove most of the bark, the areas of the tree from where the bark has been removed are pasted with a plaster of clay or mud and kept moist for a few days.

Characteristics of the Bark

Bark on young stem or branches is smooth greenish to rusty green. In transverse section, it can be differentiated into three zones:

1. A narrow pink outer part.
2. A whitish, brown middle portion.
3. Thick innermost region of secondary bast. A few secretory cavities are found scattered in the phloem.

The older bark is with numerous scattered tubercles, grey or dark grey to grey black in colour, feebly fissured and exfoliating. The entire bark is comparatively thin, about 10 mm thick. The outer cork consists of nearly half the thickness of the entire tissue. The inner surface is pink brown and fibrous. It has very feeble smell when fresh and a slight bitter taste.

In the old bark, a well-defined outer rind is present, formed of alternate strips of cork layers and dead secondary bast. The cork layer is composed of secondary cells, often with reddish brown contents. Phellogen is not very distinct and a secondary cortex is not normally present. The secondary bast is composed of groups of scelerenchyma which are polygonal, isodiametric with a thick unpitted wall and medullary rays. In still older barks, collapsed and compressed phloem tissue is present. The phloem parenchyma cells are packed with compound starch grains. Medullary rays are usually 2–5 serrate. Most of the cells contain starch. Cubical, rectangular and polyhedral crystals are quite common.

Root Bark

It has numerous oblong lenticels, 2–5 mm long, arranged regularly in longitudinal rows. The outer bark generally consists of thin walled cork cells, yellow to rusty brown in color. In older bark there may be alternate zones of thin hard crustaceous portions and soft cork tissue.

In older root in transverse section, a large number of pores and medullary rays are visible. Medullary rays are usually 2–5 serrate.

Leaves

These are collected from the tree, with the young twigs, chopped into small pieces and dried.

Characteristics of the Leaves

The detailed morphology is given in Chapter 2 on Plant Sources.

The micromorphological characters are: Leaf epidermis, adaxial cells elongated with straight or slightly undulating walls, stomata absent or a few. Abaxial cells at intercostal zone are with straight to slightly undulating walls. Stomata are common and oval in shape, anomocytic, 360–500 stomata per mm, each surrounded by 4–6 subsidiary cells and slightly raised on the general surface. In some cases stoma are plugged with thick resinous deposits. Costal cells are elongated and arranged in rows. Crystals in the form of druses, single or in groups, are present in the epidermal cells. Hairs unicellular to multicellular, 50–200 µm in length and 10–15 µm in width at the base. The leaf has a normal dicotyledonous structure, when seen in transection.

REFERENCES

Mitra, C.R. (1963) *Neem*. Indian Central Oil Seed Committee, Hyderabad, India.

6. QUALITY ASSURANCE OF PLANT RAW MATERIALS

Bark is official in the *Indian Homoeopathic Pharmacopoeia* (1971) while in the industry, neem oil and neem seed cake are used.

BARK

Indian Homoeopathic Pharmacopoeia

This describes both the macroscopical and microscopical character of the bark. As per the pharmacopoeia, the bark is dark grey to greyish black externally, while the inner surface is pinkish brown and fibrous. In old bark there is a well-defined outer rind formed of alternating strips of cork layers and dead secondary bast. The cork cells have reddish brown contents. Phellogen is not very distinct and a secondary cortex is normally not present.

SEED OIL

Azadirachtin in the samples can be estimated by the method given in Fig. 7 in Chapter 3 on chemical constituents. The physical standards of the oil are given by the Panel of the Ministry of Industry, the Government of India (Table 2), the Solvent Extractor Association of India (Table 3) and the Indian Standards Institute (Table 4).

Composition of Oil

In the earlier literature (Wealth of India, 1948), oil was reported to contain 0.2 percent myristic acid, 16.2 percent palmitic acid, 14.6 percent stearic acid, 3.4 percent archidic acid, 56.6 percent oleic acid and 9.0 percent linoleic acid. It is quite rich in tocopherols which may be up to 1.17 mg/g, having equal amounts of α and γ but very little β tocopherols. Recently Milan Mehtra (1997) has given the composition of the oil as follows: myristic acid 2.6 percent, palmitic acid 13.6 to 14.9 percent, stearic acid 14.4 to 19.15 percent, oleic acid 49.1 to 61.9 percent and linoleic acid 7.5 to 15.8 percent. The percentage of glycerides is 0.6, fully saturated glycerides 22.0 percent, tri-unsaturated glycerides 34.0 percent,

Table 2 Standard of neem oil by panel of Ministry of Industry, Government of India

Moisture and insoluble impurities = 0.9%
Saponification value = 180–205
Iodine value (Wij's) = 70–82
Unsaponifiable matter (by wt.) = maximum 2%
Flash Point (Pensky Marten Method) = minimum 120°C

Table 3 Standards of neem oil by solvent extractors association of India

	Grade I	Grade II
Moisture and insoluble impurities	0.9%	same
Color in Lovibond Scale Y + 5R in 1/4" cell	110 maximum	—
Refractive index @ 40°C	1.4615–1.4705	same
Specific Gravity @ 30°C	0.908–0.934	same
Saponification value	180–205	same
Iodine value (Wij's)	70–82	same
Unsaponifiable matter	2% max.	same
Free fatty acids, %age by wt.	12%	same
Flash Point (Pensky Marten Method)	120°C min.	same
Remarks and other characteristics	Tire 36° min. color of soap not deeper than light brownish yellow.	—

Table 4 Standards of neem oil by Indian Standards Institute

Minimum moisture and insoluble impurities (by wt.)	0.3%
Specific gravity (30°C)	0.908–0.934
Refractive index	1.4615–1.4705

stearodiolein 2.0 percent, palmitodiolein 26 percent, oleopalmitostearin 12 percent and oleodipalmein 5 percent.

SEED CAKE

Keeping in view the importance of this in agriculture, the Indian Standards Institute (Anonymous, 1977) has given specification No. 8558 for neem cake for manuring, which is as follows:

- Maximum moisture (percentage by mass) 10%
- Maximum water soluble organic nitrogen on moisture free basis (percentage by mass) 2.5%
- Maximum total ash (percentage by mass) 13%
- Maximum acid insoluble ash (percentage by mass) moisture free basis 5.0%

REFERENCES

Indian Homoeopathic Pharmacopoeia (1971) Ministry of Health, Government of India, New Delhi.
Indian Standards Institute (1968) Specification of neem kernel oil. No. 4765.
Indian Standards Institute (1977) Specification for neem cake for manuring. No. 8558.
Milan Mehtra (1997) Nicely neem. *Chemexcil Export Bulletin*, **31**, 25–28.
Solvent Extractor's Association of India, Handbook (1996) Bombay, India.
Wealth of India (1948) Council of Scientific and Industrial Research, New Delhi, India.

7. WOOD FOR FUEL AND TIMBER

Neem in recent times has emerged as a tree of choice for afforestation projects and for meeting the fuel and timber needs of under developed countries (Kalla et al., 1978; Anonymous, 1980, 1983, 1986; Radwanski and Wickens, 1981; Grainger, 1982; Prasad, 1983).

CHARACTERISTICS OF NEEM WOOD

Sapwood greyish white, heartwood red, when exposed fading to reddish brown and then resembling dull to somewhat lustrous, aromatic with characteristic taste. The wood is hard, closed grained. According to Gamble (1902), annual rings are doubtful, the wood shows alternating bands with numerous and fewer pores, and pale concentric lines but Pearson and Brown (1932) mentioned that growth rings are distinct, sharply delimited by narrow brown concentric lines. Pores scanty, moderate sized and large, often oval and sub-divided.

Medullary rays numerous, white, prominent, bent outward where they touch the pores, the distance between the rays less than the transverse diameter of the pores. Rays oval to elliptical with linear and lenticular orifice, oil globules frequent, crystals not found. Gill et al. (1985) gave the morphological characters of treachery elements, i.e. vessels, fibers, rays, parenchyma and tracheids. Vessels form inconspicuous vessel lines along the grain, which are darker than the background and contain deposits of reddish brown gum. Vessel segment short (120–450 µ) abruptly or alternate tailed on one or both sides. The perforation is simple, horizontal to oblique, frequently occupied with gum. Nair (1987) reported Scanning Electron Microscopical (SEM) studies on the helical thickening covering the entire vessel elements. Parenchyma terminal paratracheal, paratracheal zonate and meta tracheal in cambiform rows Fibers non-libriform or semi-libriform in radial rows, non-septate, 250–1620 µ in length and 22–28 µ wide.

Nair (1988) gave a revised account of wood anatomy and heartwood formation. According to the author, wood is of diffuse porous type. Axial parenchyma cells and sometimes the vessel and fiber of heartwood show the presence of extractives. The necrobiosis of the parenchyma cells occurs at the heartwood boundary. The death of parenchyma cells is associated with depletion of starch grains and accumulation of extractives.

Mechanical Properties of the Wood

The wood is moderately heavy, the average weight is 50–52 lb per cubic feet (Gamble, 1902). Pearson and Brown (1932) gave the following mechanical properties of the wood:

- Compression parallel to grains lb/sq inch = 6680
- Shear parallel to grains lb/sq inch = 1326
- Module of elasticity or Young modulus = 1,008,800

Sekhar and Gulati (1971) determined the strength parameters of neem wood with respect to teak (*Tectona grandis*) as 100, which are as follows:

- Weight at 12% moisture content = 124
- Strength and stiff as beam = 87
- Retention of shape = 77
- Shear = 129
- Surface hardness = 131
- Refractoriness = 113
- Nail and screw holding power = 117

Koul *et al.* (1990) reported the specific gravity of wood as 0.65–0.85. According to these authors, the wood is straight grained and is more resistant to shock.

WOOD AS TIMBER

After extensive study the wood was recommended for general construction (Sekhar and Gulati, 1971; Rajput and Shukla, 1984). It was found to be naturally decay resistant (Rao, 1990), absorbing a minimum amount of water so good for outdoor use (Das *et al.*, 1993) and for timber house construction (Punhani, 1995) but Kossou (1992) did not find it suitable for the construction of granaries because of susceptibility to insect attack.

On the basis of the tests for glue adhesion, tensile strength, bending, compressive strength, amenability to preservatives and fireproofing treatment, Chauhan and Bist (1987) considered neem suitable for general-purpose plywood, while Shukla *et al.* (1990) studied its carving behavior. Punhani and Pruthi (1990) measured the lateral bearing strength using nails and shear strength with joints loaded parallel, perpendicular and at 30°, 40°, 60° to the grains. Pant *et al.* (1962) found overall performance of the wood better than teak (*Tectona grandis*) as far as working properties are concerned.

The earlier literature on the wood quality of neem was studied by Tewari (1992) in detail. It has low shrinkage and seasons well in air. The wood is durable, but impregnation with chemicals like preservative and adhesives is difficult because of the gummy deposits in the cells. The quality index based on the quality performance is 114 as compared to 100 for teak. It resembles mahogany, hence is sometimes called "Indian mahogany", but lacks grain and smoothness. The compressed wood shuttle used in the textile industry, made from neem wood, was studied by Shukla and Bhatnagar (1988, 1993), but not found suitable because of lack of enough strength.

WOOD AS FUEL

Fuel wood characteristics, viz. density, calorific value, and content of C, H, N and O, were determined by Jain (1990). In Ghana, the first rotation of neem yields 108–137 m^3 of fuel wood per hectare, but in northern Nigeria 19–69 m^3 per hectare after 8 years (Tewari, 1992). Puri *et al.* (1994) determined the fuel wood value index for it, taking into account calorific value and density as positive characteristics, and high water content and high ash values as the negative points. Neem had better fuel efficiency in comparison to *Prosopis cinereria*, but was less efficient when compared with other indigenous trees like *Acacia nilotica*.

REFERENCES

Anonymous (1980) *Fire Wood Crops: Shrubs and Tree Species for Energy Production*. National Academy Press, Washington D.C., USA.

Anonymous (1983) *Firewood Crops: Shrubs and Tree Species for Energy Production*, Vol. II, *Fuelwood species of arid and semi arid regions*. National Academy Press, Washington D.C., USA.

Anonymous (1986) Silviculture of promising species for the production of fuel wood in Central America. The results of five years of investigation (in Spanish). *Informe-Technico-Centro-Agronomic Tropicale Investigacion-y-Ensenanza, Costa Rica*, No. **86**, 228.

Chauhan, B.R.S. and Bist, J.P.S. (1987) Plywood from Indian timbers: *Azadirachta indica* (neem). *Journal of the Timber Development Association of India*, **33**, 47–55.

Das, S.C., Akhter, S. and Sayeed, M. (1993) Chemical composition and water repellence property of ten village wood species. *Bangladesh Journal of Forest Science*, **22**, 61–67.

Gamble, J.S. (1902) *A Manual of Indian Timbers* (reprint 1972). Bishen Singh Mahendrapal Singh, Dehradun, India.

Gill, L.S., Lamina, B.L. and Karatela, Y.Y. (1985) Histomorphological studies of the treachery elements and the economic potential of some tropical hardwoods. *Sylvatrop*, **10**, 91–141.

Grainger, A. (1982) Firewood Crops Review of volume I. *International Tree Crops Journal*, **2**, 2.

Jain, R.K. (1990) Fuelwood characteristics of trees grown on alkaline soils of Lucknow. *Range Management and Agroforestry*, **11**, 95–98.

Kalla, J.C., Chand, G., Vyas, D.L. and Gehlot, N.S. (1978) Techno economic felling cycles for selected energy plantation species in an area of Western Rajasthan. *Annals Arid Zone*, **17**, 42–51.

Kossou, D.K. (1992) The sensitivity of wood used for the construction of traditional granaries to attack by Prostephanus truncatus (Horn) (Coleoptera: Bostrychidae). *Insect Science and its Application*, **13**, 435–439.

Koul, O., Isman, M.B. and Ketkar, C.M. (1990) Properties and uses of Neem. *Canadian Journal of Botany*, **68**, 1–11.

Nair, M.N.B. (1987) Occurrence of helical thickening on the vessel element walls of dicotyledonous woods. *Annals of Botany*, **60**, 23–32.

Nair, M.N.B. (1988) Wood anatomy and heartwood formation in neem (*Azadirachta indica* A. Juss). *Botanical Journal of the Linnean Society*, **97**, 70–90.

Pant, B.C., Shukla, K.S. and Badoni, S.P. (1992) Working qualities of Indian timbers. Part IX. *Indian Forester*, **118**, 573–582.

Pearson, R.S. and Brown, H.P. (1932) *Commercial Timbers of India*. Vol. I. Government of India, Central Publication Bureau, Calcutta, India.

Prasad, R. (1983) Neem—the miracle tree for meeting Indian's growing energy needs. *Bio-energy News*, **1**, 43–47.

Punhani, R.K. (1995) Span tables for roof pulins and ceiling joints in some social forestry species for timber house constructions. *Indian Forester*, **121**, 651–662.

Punhani, R.K. and Pruthi, K.S. (1990) Lateral bearing strength of *Azadirachta indica* (neem) under common wire nails. *Indian Forester*, **116**, 825–831.

Puri, S., Singh, S. and Bharatbhushan (1994) Fuel wood value index in components of ten species of arid region in India. *Industrial Crops and Products*, **3**, 69–74.

Radwanski, S. and Wickens, G.E. (1981) Vegetative fallows and potential value of neem tree (*Azadirachta indica*) in the tropics. *Economic Botany*, **35**, 398–414.

Rajput, S.S. and Shukla, N.K. (1984) *A Resume of the Classification of Timbers for Different End Use*. Controller of Publications, New Delhi (India).

Rao, R.V. (1990) Natural decay resistant of neem wood. *Journal of the Indian Academy of Wood Science*, **21**, 19–21.

Sekhar, A.C. and Gulati, A.S. (1971) Suitability indices of Indian timbers for industrial and engineering uses. *Indian Forest Record*, 2, (*vide* Tewari D.N. (1992) *Monograph on Neem* (*Azadirachta indica A. Juss.*). International Book Distributors, Dehradun, India.

Shukla, K.S. and Bhatnagar, R.C. (1988) Suitability indices of Indian timbers for compressed wood shuttle blocks. *Journal of Timber Development Association of India*, **34**, 53–58.

Shukla, K.S. and Bhatnagar, R.C. (1993) Compressed wood from some special forestry species. *Journal of the Timber Development Association of India*, **39**, 5–8.

Shukla, K.S., Pandey, K.N., Pant, B.C. and Bedoni, S.P. (1990) Carving behavior of some Indian timbers—quantitative approach. *Journal of the Indian Academy of Wood Science*, **21**, 27–32.

Tewari, D.N. (1992) *Monograph on Neem* (*Azadirachta indica A. Juss.*). International Book Distributors, Dehradun, India.

8. PROCESSING OF PLANT RAW MATERIAL

THE SEED

According to an estimate, there are 13–15 million neem trees in India alone. If one tree produces, say, 50 kg of seed and even a small percentage of the total yield is processed for the oil, the quantity of oil and seed cake so produced would be enormous. The industry based on neem (Fig. 10) will not only provide extra income to the rural poor for the collection of seed, but produce a fatty oil, which can be raw material for many industries. As a by-product, neem seed cake will be available for use directly as organic manure and indirectly as a nitrification inhibitor and pesticide.

Neem as a source of oil has been of interest to both social workers and scientists for a very long time in India, but the yield of oil by traditional methods was less; moreover, the oil obtained by bullock-driven expellers, called "*Kohlu*" or sometimes "*Katchi Ghani*", was darker in color, fetid in odor and bitter in taste. It was not acceptable to consumers in any form and could not be put to any major industrial use. It was used by the lower strata of society to a limited extent for the skin diseases of man and farm animals, as a lubricant for farm implements, particularly for the leather straps, for prolonging their life and as an oil for use in clay lamps for illumination.

The earlier studies were confined to refining the oil by chemical means by removing the bitter principles. Sen and Banerjee (1931) studied these bitter principles. Child and Nathanael (1944) gave the fatty oil composition of neem, which was very similar to the other edible oils. Siddiqui (1942) presented a note on the isolation of three new bitter principles from neem oil, while Siddiqui and Mitra (1945) suggested various ways by which these principles could be utilized in the pharmaceutical industry. Quite a number of patents were also taken out on this aspect (Siddiqui and Mitra, 1945a, 1945b, 1945c), but these studies could not lead to the industrial utilization of neem oil, because the oil

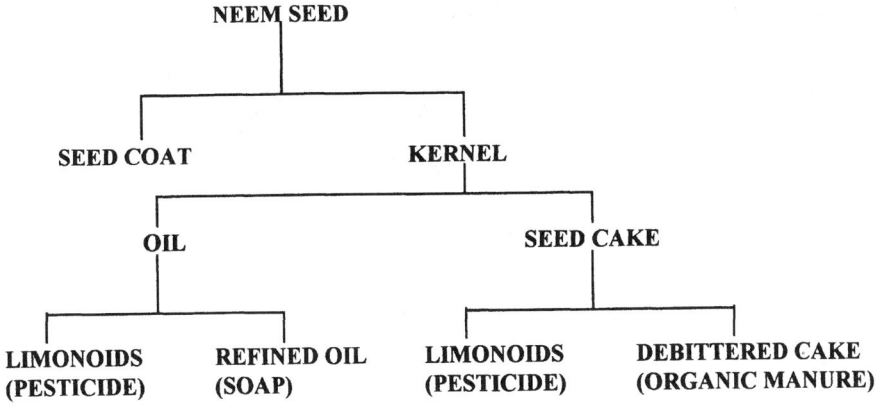

Figure 10 Flow diagram to show the major products which can be obtained from neem seed

obtained was not of a desirable quality even after refining, and there was no major use to which the bitters could be put.

Collection

When the problem of not getting good quality oil even after refining was studied in detail, it was seen that it was mainly due to the low quality of the raw material used. The oil made from these heavily contaminated low quality seed was resinous and malodorous. The Indian Oil Seed Committee, then located at Hyderabad, looked into all the above aspects and Mitra (1963) came out with an excellent book, *Neem*, in which on the basis of field studies, experiments and trials, methods of obtaining a better quality of oil by drying the seed properly before storage were given. Mitra (1979) gave an account of technology, incorporating the other developments.

Decortication

The seeds are dried once more in dry air or in dryers and are subjected to decortication, i.e. removal of the kernel from the seed coat. It can be done in various ways. Mitra (1963) suggested double roller mills, with rollers moving in opposite directions. He used cast iron rollers with various types of grooves. These days, the rollers have been replaced by decorticators used for dehusking peanuts. These decorticators essentially consist of rubber rollers with a blower at the base of the chamber. The roller presses the seed coat and releases the seed kernel; the mixture of both seed and seed coat passes through a sieve into the chamber, where the lighter seed coat gets blown away, keeping the seed kernel in the chamber.

In improved decorticators (Figure 11B) now, a single roller rotates above a sieve, which presses the seed between them, causing separation of the seed coat and the kernel.

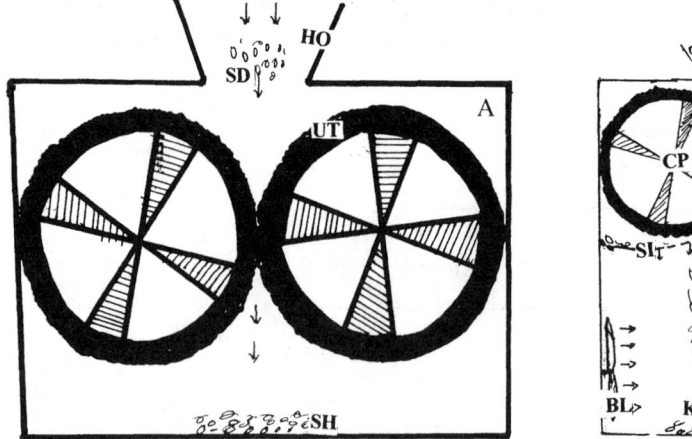
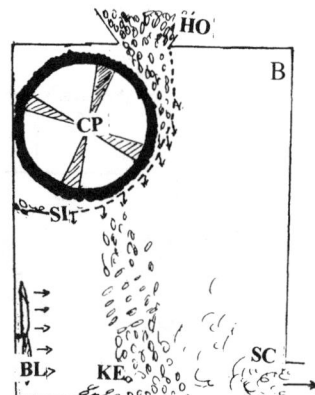

Figure 11 Decorticators. A, an improvised technique using old tyres; B, flow diagram of improved machine. *Abbreviations*: BL = blower, CP = central pulley, HO = hopper, KE = kernel, SC = seed coat, SD = whole seed, SH = seed husk, SI = sieve, UT = used tyres

On a small scale, the seed coat can be separated by taking two old car tyres (Fig. 11A). These are fixed by a central shaft so that the two tyres are adjacent to each other. The whole seed is fed between the gaps of the tyres, which get pressed by their movement, causing the separation of the seed from the outer cover; the two can be separated from the mixture by a winnower. This winnower in earlier times was made from bamboo strips or reeds. It essentially consists of a broad plate, tapering at the broad end, and having a raised platform on the other. This winnower is moved left to right and up and down, after loading with the mixture, and by this action the heavy seed remain on the upper part of the winnower while the lighter chaff comes out at the open free end. An improvised winnower can now be easily made by cutting a big plastic jerry can obliquely. The plate so obtained has a raised platform on one side, raised sides and a tapering free end.

Winnowing can also be done by utilizing the wind direction. The mixture of chaff and seed is placed on the winnower and is lifted by a person in a standing position. The winnower is shaken slowly by hand movement in the direction of the wind so that the mixture is allowed to fall on the ground. By this action the lighter chaff flies away because of the wind and the heavy seed kernel falls on the ground, nearly perpendicular to the standing person. This operation can also be performed by using a blower.

The separated seed kernels so obtained can be used for oil extraction.

Extraction of Oil

Traditional Method

Oil extraction in India was a specialized profession and the people undertaking this job were called "*Teli*" or the oilmen, and the traditional oil extractor was called "*Kohlu*" (Fig. 12A). One well-known version of it, available now, is called "*Katchi Ghani*" or the cold process, in which heat is neither applied nor generated (Fig. 12B).

In earlier times, the dried whole fruit with seed coat intact was used for extraction in a *kohlu* at certain moisture levels, and no efforts were made to remove the seed coat or pulp, but the yield of oil was very low because the seed coat absorbed most of the oil liberated from the kernel.

The *Kohlu* was made of wood in earlier days. It essentially consisted of a pestle and mortar connected to wooden plank. The pestle had a metallic cover at the base and the inner surface of the mortar was made rough with strips of wood or bamboo. For the extraction of oil, the seed were fed into the mortar. The wooden plank was rotated by blind folded oxen in a circular motion. This caused the movement of the pestle along the sides of the mortar, crushing the seed which came in between the two. Water was sprinkled on the seed from time to time to maintain a particular level of humidity. The oil, on extraction, settled down at the base of the mortar, and was collected by a bamboo pipe, inserted in a hole at the base of the mortar.

The present modified version of the above *kohlu* (Fig. 13A) consists of a rotating mortar installed on an electric motor. The pestle is stationary, and can be adjusted along the sides of the mortar by a spring and rod adjacent to it. After feeding the seed, the mortar is rotated, and the seed which come in between the pestle and the mortar get crushed, liberating the oil. Because of the low yield of oil, the use of above cold

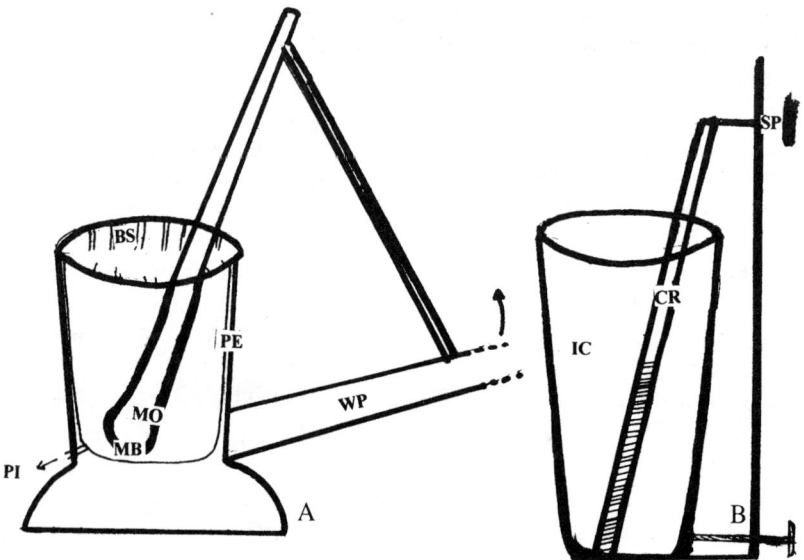

Figure 12 Extraction of oil by cold process (*kachi ghani*). A, traditional machine; B, improved version of the traditional machine. *Abbreviations*: BS = bamboo stips, CR = central rod, IC = iron container, MB = metallic base, MO = mortar, PE = pestle, PI = bamboo pipe, SP = spring to adjust the central rod, WP = wooden plank

process was discouraged by Indian government agencies and stress was laid on solvent extraction plants or on big mechanical expellers.

Much of the neem oil in small-scale industry is obtained by expellers made of iron, which is a screw-like press in a horizontal position. In improved versions, a boiling kettle is attached to the expeller which is heated by steam. The yield of oil is greater when using these mechanical devices, but the quality of oil is very poor, because of metallic contamination and the use of excessive heat. The oil obtained is black, tar-like, resinous and with a bad odor. The steam required by the industry is generated by waste-fired boilers, where neem seed coat is used as a fuel.

Since the seed coat is not an efficient fuel and has low combustibility, it has to be fed regularly to the furnace with a spade. To avoid the seed coat forming lumps, a staircase like structure is placed at the mouth of the furnace so that the seed coat gets well spread on the fire inside the furnace. If the seed coat is fed in a mass onto the fire, it extinguishes it.

Filtration of Oil

The oil obtained by traditional methods has an appreciable amount of suspended particles, which do not get filtered by coarse cloth or by simple filtration techniques. The best method for their removal is to let them settle by keeping the oil undisturbed for a few days. After that, the oil can be decanted off. The sediment which has settled at the bottom is very rich in oil and can be used as such in place of oil or can be re-fed to the expeller along with the seed for the extraction of oil.

When the quantity of the oil to be filtered is large, the oil is subjected to a mechanical filter press. This filter press essentially consists of wooden or iron frames, and a long filter cloth (Fig. 13B). With the help of a motor, when the frames are pressed along with the filter cloth, the oil gets filtered. It is a common practice to filter the oil twice, and the oil so obtained is sold as "double filtered".

If a very refined oil is required, the oil may be processed in a centrifuge or a filter aid may be used.

Solvent Extraction

Much of the oil in industry is obtained by solvent extraction, using hexane as a solvent. It is based on the principle of Soxhlet extraction, in which the seed is repeatedly extracted with the same solvent. In Fig. 14, a photograph of a pilot plant is given. Solvent extraction gives maximum yield of good quality oil.

In the case of neem, the starting material may be the whole kernel or the seed cake obtained after the cold process or from expellers. In this process, the seed or seed cake is crushed to a particular particle size and mixed with hexane which is used as a solvent. The oil dissolved in the solvent is filtered to leave the hexane-insoluble portion in the sieve. The mixture of hexane and oil is subjected to distillation which separates the volatile hexane from fatty oils. The solvent residue from this oil may be removed by passing steam through it or by de-odorization.

Because of the hazardous nature of hexane, some alternative solvents such as propyl alcohol have been suggested, but they are uneconomical.

Refining of Oil

In spite of the best efforts by chemists in India, neem oil free of bitter principles could not be obtained. This is because of the unique nature of these principles. These are so

Figure 13 Views of an oil expeller unit. A, expeller; B, filter press

Figure 14 A pilot plant for solvent extraction. The neem seed are fed into the vessel, extracted with petroleum ether, hexane or any other suitable solvent, filtered and the filtrate is subjected to solvent recovery, to obtain fatty oils along with fractions effective as pesticide (Photo by author)

intimately mixed with oil that their separation during the refining of oil was not possible for a very long time. Since 1945, all the well-known methods for refining oil have been tried, but the results were not encouraging. Ahuja *et al.* (1976) worked on miscella refining of oil. Ramachanderaiah *et al.* (1977) tried alcohol segregation. In one alternative process, neem oil was saponified and grained, while in the second process boiling was carried out after dilution with water and adding a higher concentration of caustic lye. The odoriferous compounds along with the free fatty acids were removed by distillative de-acidification or by splitting into odorless fatty acids (Milan Mehtra, 1997).

The Central Oil Technology Research Institute in India has recently developed a process for the cold extraction of neem oil, rich in 2000 ppm or more of azadirachtin (Ramkrishna *et al.*, 1995). In this process, the temperature during the whole operation does not go beyond 40°–45°C.

An account of an improved technology for the extraction of oil and for the recovery of azadirachtin has been given by Waghray (1993), in which seed collected from low humidity areas are immediately de-pulped and dried in a continuous drier. Double twin-roller decorticators are utilized with continuous winnowing for the separation of the kernel, and from it oil is extracted. The whole process is monitored throughout by High Performance Liquid Chromatography for azadirachtin.

Standards of the Oil

Keeping the economic importance in view, particularly for the soap industry, standards for neem oil have been developed by various organizations; these are given in Tables II, III and IV of Chapter 6.

PESTICIDE FORMULATIONS

Commercial

Mainly two types of neem-based products are now available. In the first case, a standardized quantity of azadirachtin is incorporated, along with other compounds, to make an azadirachtin-rich product having a long shelf life. In the other case, azadirachtin-rich neem oil is taken and fortified further with azadirachtin to a particular level. In the first case, pesticidal activity is due only to azadirachtin and there is every possibility that some of the insects may develop resistance to it. In the second case, in addition to azadirachtin the formulation has many compounds, naturally present in neem oil. This type of product may have a multi-pronged attack and even if the pests develop resistance to one active ingredient, there are other compounds in it for pesticidal effect.

On a Small Scale

In Wealth of India (1948), several recipes for the control of pests by small-scale farmers are given using neem-based formulations. All of these were developed before the unstable nature of the active constituent azadirachtin was known, and might not have been very effective. The simple formulations were painting the tree trunk with neem oil, or preparing a water in oil emulsion for spraying on crops, using household detergent or mixing a paste of neem kernel prepared with a stone grinder in water and filtering the emulsion through a coarse cloth. Lange and Feuerhake (1984) conducted experiments to see if synergist-like piperonyl butoxide could be used in this type of preparation. Rajasekaran and Kumaraswami (1985) studied other methods by which the efficacy of these extracts could be increased. Schmutterer (1985) successfully demonstrated the suitability and application of neem in the control of pests in the tropics.

From a crude neem kernel extract, Feuerhake and Schmutterer (1985) developed an inexpensive standardized formulation for small-scale farmers by using crude neem kernel extract, solvent glycol ether and propyl gallate benzophenone. The activity was confirmed by bio-assay and chromatography.

Ketkar (1989) in the Neem Mission in India propagated several simple neem preparations for the use of small farmers, but in the absence of any direct anti-knocking agent of neem formulations as compared to synthetic chemical ones, these preparations did not become popular. It was difficult to convince a peasant that a common tree like neem can be effective against insects.

Olaifa et al. (1993) were aware of the unstable nature of azadirachtin, so they developed a technology to improve the shelf life of an emulsifiable concentrate from neem oil by using an aqueous extract of *Tetraplema tetrapera* L., along with 0.1 percent propyl paraben and 0.1 percent octylgallate, which prevented microbial spoilage and oxidation of emulsion. Optiz (1991) developed methods which could be utilized by small-scale farmers, by using their own skills.

Vijayalakshmi et al. (1995) described a number of methods which a villager could apply without the use of modern technology, but incorporating modern knowledge about the degradability of azadirachtin in the environmental conditions. Heating was to be avoided in all cases; emulsion should be freshly prepared, and sprayed in the morning and evening, when there is very little sunshine. In summer the frequency of spraying should be greater. Spray could be prepared by soaking neem seed, leaves, etc. overnight and straining through a thick cloth in the morning. In the case of oily preparations, common household detergent may be added as an emulsifier. These authors did not mention the use of soap prepared from neem oil for spray, but this could prove very helpful for the non-cereals crop maturing in March–April, where attack by aphids is so wide-spread, particularly in mustard fields, that these insects could be seen suspended in air like dust particles, even in urban areas.

For Treatment of Bags

It is common practice in India to store the grains in used bags for economic reasons. These bags, or even new ones, harbor the worms or their eggs before use and attack the cereals stored in them. Much of the infestation of these grains by well-known worms can be prevented if these bags are treated with some pesticide before use. Vijayalakshmi et al. (1995) have recommended the soaking of jute bags in 10 percent neem kernel emulsion for 15 minutes and then drying them, before use.

In the villages, it is common practice to plaster hut or floor or small storage bins with clay or mud frequently. For this mud plastering, neem seed or oil emulsion can be incorporated in the mud mixture in place of water to keep the pests away.

NEEM OIL AND SOAP TECHNOLOGY

Before synthetic detergents became popular in India, soap making, particularly for laundry, was a cottage industry and any readily available, cheap fatty oil was incorporated in soap formulations. Neem oil was one of the ingredients, but was not preferred because of the malodor and dark color of the soap so obtained. Conventional industrial methods failed to refine the oil because the bitter principles are mixed with the lipid constituents so intimately that it is difficult to separate the two. Mitra (1963) gave an account of various steps taken for refining neem oil for the soap industry. In due course of time, the soap manufacturers learnt that neem oil could be incorporated up to 15 percent into the mixture of fatty oils along with cotton seed oil, coconut oil, etc. The cotton seed oil was found to mask the odor of neem.

For the manufacture of soap on a small scale by the cold process, vegetable oils are mixed with 13 percent caustic flakes and 25 percent water and stirred vigorously until a thick paste is formed, which on keeping overnight becomes solidified and is cut into cakes. Ketkar and Ketkar (1995) have also provided information about other formulations for this type of soap making.

In an alternative method, on a small scale, neem oil with an excess of water and caustic soda is heated and after boiling, allowed to cool. Many of the impurities either escape in the air by heating or get dissolved in water. This mixture of neem oil and caustic soda is further treated with other fatty oil to get a soap of the desired composition.

In large-scale operation, for making toilet soap, one method suggested was to prepare noodles as above, while the other method consisted of the separation of fatty acids from crude neem oil. These pure fatty acids are then reacted with alkali to get soap. Milan Mehtra (1997) gave a method for the production of quality soap by saponifying the neem oil and graining it, where most of the smell and the color are separated out. A second boiling is then carried out by diluting with water and adding a higher concentration of caustic lye. The author has also suggested distillative de-acidification. The distilled odorless oil so obtained is subjected to the usual splitting and distillation, to get a light-colored odorless soap.

NEEM OIL IN THE LEATHER INDUSTRY

Common salt (sodium chloride) is commonly used for curing hides and skin, and is a major pollutant in tanning effluents. In a process developed by the Central Leather Research Institute in India, neem oil has been used in place of salt with good results (personal communication).

NEEM OIL AS FUEL

Neem oil as a substitute for petroleum-derived fuel has been considered from time to time. Mitra (1963) has given an account of all these developments. By pyrolytic degradation a product, "pyronimin," was obtained, which could be converted into many industrial products. Munavu (1984) considered it a non-conventional vegetable oil for fuel. Bansal and Juneja (1989) studied the performance of neem oil as a diesel engine supplement fuel, while Renuka and Ramani (1989) tried neem cake as a source of bio-gas.

OTHER USES

Neem can be used for the production of olein and high melting stearin after hydrogenation. Kane and Kulkarni (1954) carried out high pressure hydrogenation of oil. A process for edible neem oil has been developed by Lidert (1994) in which crude oil is treated with an alkaline solution or hydrogen peroxide and subjected to distillation or chromatography. The oil so produced has a very low sulfur content.

NEEM CAKE AS CATTLE AND POULTRY FEED

Neem cake is very rich in proteins and comparable to soya bean, peanut or even fish meal, and can be partly substituted for any of these in cattle and poultry feed (Ketkar, 1995). Neem seed cake is very rich in minerals and is easily converted into organic minerals by animals during digestion. If neem cake is supplemented in the diet, the animals do not require additional inorganic minerals in the diet. It is very rich in amino acids like methionine and lysine. Further details are given in Chapter 14.

Leaf extract can be used as a preservative in food to prevent it from bacterial infestation and aflatoxicosis. Neem oil in very low concentration may be used as an antimold for poultry and cattle feed.

RECENT DEVELOPMENTS

Earlier, the stress was on obtaining an odorless neem oil for non-edible purposes, but with the discovery of azadirachtin, processes have now been developed for the isolation of bitter principles, and a good quality oil is obtained as a by-product.

For the recovery of pesticide, Amata-Archachai and Wasuwanich (1986) suggested a method for the de-pulping of seed, storage and extraction. According to Hoschle-Zeledon and Keyserlingk (1993), a pilot plant for the production of a standardized extract of neem as an insecticide has been in operation in Myanmar (Burma) since 1987. In this process, the seed are extracted with methanol to get the crude extract rich in azadirachtin. The extract is formulated with an emulsifier and diluted with methanol. Waghray (1993) described a new technology for pesticides and oil. Ramakrishna *et al.* (1995) have developed a cold process for neem oil. Sivakumar *et al.* (1995) have described various machines for processing neem, while Dakshinamurthy (1995) has given an account of technology for the production of azadirachtin.

REFERENCES

Ahuja, M.M., Gupta, R.R., Agarwal, K.N. and Gupta, A.C. (1976) Miscella refining of neem seed and mahua oils. *Indian Journal of Technology*, **14**, 257–259.

Amata-Archachai and Wasuwanich, P. (1986) Mechanical extraction and cleaning of nuts of some tropical species. *Embryon*, **2**, 1–8.

Bansal, B.B. and Juneja, N.N. (1989) Performance evaluation of neem oil (Melia azadirachta) as diesel engine supplementary fuel. *Agricultural Engineering*, Dublin, Ireland, 4–8 Sept., 1989.

Child, R. and Nathanael, W.R.N. (1944) On the fatty acids of margosa (neem) oil. *Journal Indian Chemical Society*, **21**, 35–37.

Dakshinamurthy, A. (1995) Technology for production of insecticides of plant origin at rural level. In R.P Singh, M.S. Chari, A.K Raheja and W. Kraus (eds.) *Neem and Environment* Volume 1–2, International Book Distributor, Dehradun (India).

Feuerhake, K. and Schmutterer, H. (1985) Development of a standardized and formulated insecticide from a crude neem kernel extract (in German). *Zeitschrif fur Pflanzenkrankheiten und Pflanzenschutz*, **92**, 643–649.

Hoschle-Zeledon, I. and Keyserlingk, N.V. (1993) Production of a standardized insecticide based on neem in a pilot plant in Myanmar (abstr.). *World Neem Conf.*, Bangalore (India), 24–28 Feb. 1993.

Kane, J.G. and Kulkarni, K.B. (1954) High pressure hydrogenation of some inedible oils and fats. *Journal of Scientific and Industrial Research*, **13B**, 890.

Ketkar, C.M. (1989) Neem (*Azadirachta indica*), as an ecologically safer potential insecticide for agricultural crops. *Changing Villages*, **8**, 1–10.

Ketkar, C.M. and Ketkar, M.S. (1995) Neem seed crush and deoiled cake as manure and as nitrification inhibitor. In Schmutterer, H. (ed.) *Neem tree, source of natural products for integrated pest management and medicinal, industrial and other purposes.* V.C.H. Verlag. Weinhim, Germany, pp. 531–540.

Ketkar, C.M. and Ketkar, M.S. (1995) Soap production from mixture of neem oil with other non-edible or edible oil. In Schmutterer, H. (ed.) *Neem tree, source of natural products for integrated pest management and medicinal, industrial and other purposes.* V.C.H. Verlag. Weinhim, Germany, pp. 549.

Lange, W. and Feuerhake, K. (1984) Increased effectiveness of enriched neem seed extracts by the synergist piperonyl butoxide under laboratory conditions.(in German) *Zeitscheift fur Ange-mandte Entomologie*, **98**, 368–375.

Lidert, Z. (1994) *Edible neem oil.* US Patent No 5371254. Rohm & Hauss Co., USA.

Milan Mehtra (1997) Nicely neem. *Chemxcil Export Bulletin*, **31**, 25–28.

Mitra, C.R. (1963) *Neem.* Indian Central Oil Seed Committee, Hyderabad, India.

Mitra, C.R. (1979) Waste fruits and seeds: Chemistry, Technology and utilization. *Progress in Plant Research*, **2**, 165–183.

Munavu, R.M. (1984) Non-conventional vegetable oils for fuel in Kenya. Bioenergy. In H. Egneus and A. Ellegard (eds.) Volume II *Biomass resources*, Elsevier Publishers, Barking, UK.

Olaifa, J.I., Orafidiya, O.O., Faniran, O.O. and Adenuga, A.O. (1993) Formulation of a locust lotion concentrate for grass hopper control (abstr.) *World Neem Conf.*, Bangalore. 24–28 Feb. 1993.

Optiz, M. (1991) Extraction of natural insecticides for use by small scale farmers as a source of income (in German). *Entwicklung Landlicher-Raum*, **5**, 22–25.

Rajasekaran, B. and Kumaraswami, T. (1985) Studies on increasing efficacy on neem seed kernel extract. In A. Regupathy and S. Jayaraj (eds.) *Behavioural and physiological Approaches in Pest Management.* Tamilnadu Agricultural University, Coimbatore, India, pp. 29–30.

Ramachanderaiah, O.S., Lakshminaryana, T., Azeemoddin, G., Ramaya, A.D. and Thirumala Rao, S.D. (1977) Alcohol segregation and refining of neem oil. *Indian Chemical Journal*, **11**, 1–4.

Ramakrishna, G., Prasad, N.B.L. and Azeemoddin, G. (1995) Cold Processing of Neem seed. In R.P. Singh, M.S Chari,. A.K Raheja and W Kraus (eds.) *Neem and Environment*, International Book Distributors, Dehradun, pp. 931–940.

Renuka, S. and Ramani, M. (1989) Castor and neem cake plus willow dust using bio gas plant: Economic feasibility and fuel efficiency. *Khadi Gramodyog*, **35**, 206–210.

Schmutterer, H. (1985) Trials of biological and integrated pest control in the tropics. *Giessene -Beitrage zur Entwicklungsforschung I Symposien*, **12**, 143–150.

Sen, R.N. and Banerjee, G. (1931) The bitter principle of neem oil. Part I. *Journal Indian Chemical Society*, **8**, 773–776.

Siddiqui, S. (1942) A note on the isolation of three new bitter principles from the neem oil. *Current Science*, **11**, 278.

Siddiqui, S. and Mitra, C. (1945) Utilization of Neem oil and its bitter constituents (nimbidin) series in pharmaceutical industry. *Journal of Scientific and Industrial Research*, **4**, 5.

Siddiqui, S. and Mitra, C. (1945a) Isolation of new bitter principles from neem oil. Indian Patent No. **33640**.

Siddiqui, S. and Mitra, C. (1945b) Separation of the bitter principles of neem oil and simultaneous refining of oil. Indian Patent No. **33650**.

Siddiqui, S. and Mitra, C. (1945c) Manufacture of new derivatives or subsidiary products from the physiologically active bitter principles of neem oil. Indian Patent No. **33651**.

Sivakumar, S.S., Palanisamy, P.T., Varadharaju, N., Gothandapani, L. and Swaminathan, K.R. (1995) Machineries for neem processing. In R.P. Singh, M.S. Chari, A.K. Raheja and W. Krauss (eds.) *Neem and Evnironment*. Volume 1–2, International Book Distributors, Dehradun, India, pp. 909–920.

Stark, J.D. and Walter, J.F. (1995) Neem oil and neem oil component affect the efficacy of commercial neem insecticides. *Journal of Agricultural and Food Chemistry*, **43**, 507–512.

Vijaylakshmi, K., Radha, K.S. and Vandana Shiva (1995) *Neem: A User's Manual*. Research Foundation for Science, Technology and Natural Resource Policy, New Delhi, India.

Waghray, A.P. (1993) Industrial technology for producing neem bitter concentrate (abstr.). *Proc. World Neem Conf.*, Bangalore.(India), 24–28 Feb. 1993.

Wealth of India (1948) Vol. I, Publication and Information Directorate, CSIR, New Delhi, India.

9. TRADITIONAL USES

The therapeutic efficacy of neem must have been known to man since antiquity as a result of constant experimentation with nature. Ancient man observed the unique features of this tree: a bitter taste, non-poisonous to man, but deleterious to lower forms of life. This might have resulted in its use as a medicine in various cultures, particularly in the Indian subcontinent and later on in other parts of the world.

AYURVEDA

The word neem is derived from Sanskrit *Nimba*, which means "to bestow health"; the various Sanskrit synonyms of neem signify the pharmacological and therapeutic effects of the tree. It has been nicknamed *Neta*—a leader of medicinal plants, *Pichumarda*—anti-leprotic, *Ravisambha*—sun ray-like effects in providing health, *Arishta*—resistant to insects, *Sheetal*—cooling (cools the human system by giving relief in diseases caused by hotness, such as skin diseases and fevers), and *Krimighana*—anthelmintic. It was considered light in digestion, hot in effect, cold in property.

In earlier times, patients with incurable diseases were advised to make neem their way of life. They were to spend most of the day under the shade of this tree. They were to drink infusions of various parts of the tree or stem sap, if available, when thirsty, eat tender leaves as salad and cooked leaves as a vegetable. Young twigs were to be used for oral hygiene and gum as lozenges for dryness of the throat and to allay thirst. Whenever mature, ripe fruits were available, they were to be sucked for their sweetish tasty pulp.

Occasionally the seed were also swallowed. While on neem therapy, patients were to avoid all products of animal origin, such as egg, flesh, milk and alcohol, which were considered "hot" in nature. It was considered that the diseases for which neem was specified were caused by hot conditions prevailing in the body, and they were mitigated by the cooling effect of neem and other herbs.

The above observations led to the widespread use of neem, from the era of Carak Samhita (200 BC–200 AD) to recent times in the Indian subcontinent and adjoining countries. It became so popular in ancient India that some scholars believe that it was an ingredient of up to fifty percent of Ayurvedic preparations. Physicians at that time advised *Panchang* of neem, i.e. five parts of the tree, the leaves, bark, fruit, flower and root. With further development in Indian medicine, the raw single-plant-part therapy was replaced by compound preparations. In these, many ingredients were incorporated; some had a complementary effect, some had a supplementary effect, some were considered antagonistic to the deleterious effects of certain herbs and some were nutritive.

Neem preparations were prescribed in various forms: *Churn*—powder, *Kwath*—decoction, *Khand*—mixed with sugar, *Kshara*—alkali, obtained by burning the plant part, *Vatika*—pills, *Asav* and *Arishta*—fermented decoctions, *Ghritam*—butter fat extract, *Malham*—ointment, *Dhupan*—fumigant, *Lep*—poultice and *Nasayam*—nasal drops.

Ayurvedic Pharmacopoeial Products

Some of the important polyherbal neem preparations of the Ayurvedic Pharmacopoeia and their main uses are:

Aparjith Dhup—fumigant for purification of air (air sterilizer)
Brhamanisthadih kwath—skin diseases
Dhattur tailam—oil for skin diseases and muscular pain
Jatyadi tailam—oil for ulcer
Jeevanti adi Kashyam—for smallpox
Laghu Manjishtadi kwath—decoction for skin diseases
Kandavadu Lepah—poultice for itching
Maha tikatam ghritam—butter fat for skin diseases
Maha tikatam kashyam—a bitter tonic
Naryana tailam—oil for rheumatic disorders
Nimbadi Kashyam—for skin diseases
Nimbadi jatailam palit—for baldness
Palit Nasyam—nasal drops for alopecia
Panch titktam ghritam—butter fat for latent fevers
Panch nimba churnam—powder for skin diseases
Patoladi kvath—decoction for fevers
Phaladi kvath—decoction for expelling worms
Punravadi kshyam—for skin diseases
Punravadi kvath—decoctions for swellings
Sudarshan churnam—powder for fevers
Thiktakam ghritam—butter fat for skin diseases
Thiktakam kashyam—for skin diseases
Varanejatayidi ghritam—butter fat for ulcers
Yograjaguggulu—for rheumatoid arthritis

All the above are polyherbal compounds, having many herbs and a special method of preparation in each case, which may not be of interest to all readers. As an example, details of *Nimbadi kashyam* are given here to give some idea of these products. This is prepared by making a decoction of 50 gm each of *Azadirachta indica*, *Tinospora cordifolia*, *Zingiber officinale*, *Curcuma longa*, *Adhatoda vasica*, *Trichosanthes dioica*, *Solanum indicum*, *Terminalia chebula*, *T. belerica*, and *Phyllanthus emblica*. This herbal mixture is boiled in 8 L of water, until reduced to 1L. Usually 60 ml of this decoction is taken twice daily on an empty stomach.

As an Antimalarial

In the nineteenth century, European physicians practicing in India and Indian practitioners of the orthodox system of medicine (allopathy) found neem bark an effective therapeutic agent for fevers, particularly for malaria, which was very common in some parts with a tropical or sub-tropical climate. The powdered bark and fresh leaves were made official in the *Pharmacopoeia of India* (Dey, 1896). It was also included in

Practical Materia Medica (Clarke, 1900). In due course of time, during the first world war or thereabout, cinchona or salts of quinine were introduced into India from England but these could not reach most of Indian population living in remote areas. Neem bark was tried as a substitute for cinchona; it did not have any direct effect on the malarial parasite, yet patients obtained relief in most cases.

The use of a decoction of bark as an antipyretic is well known, particularly for malaria. For making the decoction, 15 gm of bark should be boiled in fifty times that amount of water, until it is reduced to 50 ml. It is strained and then 10 ml of this filtrate is given thrice daily.

In Venereal Diseases

Before the arrival of the Portuguese and other European colonizers, venereal diseases were not known in India. These diseases, particularly syphilis, were noticed in some cosmopolitan areas but local physicians could not diagnose them as there was no mention of the symptoms of these diseases in their texts and they nicknamed them "*Firang rog*", or foreign diseases. As per the concept prevailing at that time, the evident symptoms were due to *hotness* in the body and *impure blood*, as for other skin diseases like leprosy, eczema, leucoderma, etc. and they prescribed neem as a *cooling agent* and a *blood purifier*.

In Ayurveda, as mentioned earlier, use of all five parts of the tree together was considered the best, but later, on the basis of experience, physicians used one or two parts together, so details of each part are given separately here. Recently Lok Parampara Samvardhan Samithi has given a data base for neem, and the interested reader may consult the data base, in addition to the following details.

Uses of Plant Part

Bark

When peeled it has two different zones; the outer one is dry, scaly, darker in color, and the inner one is smooth and brown in color. For medicinal purposes, the inner portion of the bark should be used, preferably when it is fresh. Root bark is considered better, but is now very difficult to get, because of the damage that may be caused to the tree during collection. The outer portion of the bark is rich in tannins and is astringent, whereas the inner region is rich in secondary metabolites. The inner bark contains a bitter principle of a resinous nature, which when moistened emits the smell of sulfur compounds like those found in garlic.

Bark exerts a strong antimicrobial and astringent effect due to the presence of phenolic compounds and tannins, which have a strong healing effect on the skin. For this reason it is an ingredient of preparations for pimples, piles, wounds, bleeding gums, etc.

In the case of pimples, a decoction of bark with the pod of *Cassia fistula* is applied, while for piles, 3 gm of bark powder with 5 gm of cane sugar is administered three times a day and neem oil is applied externally. A tooth powder made from a fine powder of bark and alum (sodium aluminium sulfate) is used for spongy gums, particularly when they are bleeding. This mixture also has a good antiinflammatory effect.

The decoction of bark was also prescribed for plague.

In the nineteenth century, European physicians in India considered neem bark to be a bitter tonic and prescribed it in place of gentian or quassia, in the form of an infusion or tincture.

Sap from the Tree

It is said that after the tree reaches a hundred years of age, on a day which cannot be predicted, it begins to exudate a nectar or sap from the crevices of the bark. The sap is thick, sweet in taste and fetid in smell. Great virtues are ascribed to this sap and it is said to be a panacea, particularly for leprotic ulcers and skin diseases of various etiology, particularly those which, as per the Ayurvedic concept, are associated with heat in the body.

Gum

The gum is very much like gum acacia in physico-chemical properties but is darker in color. It is a demulcent and is used for sore throats.

Wood

A pestle and mortar made from neem wood is preferred for pounding herbs. The wood is rubbed on a wet stone to form a paste. This paste is used for dressing wounds.

Leaves

The tender leaves were cooked like spinach in ancient India and sometimes fried in butter oil (the fat obtained after heating butter and discarding the fat insoluble part). To remove the bitterness from the leaves, they were boiled in water, and the leaves so obtained were cooked with some sour fruit like those of *Embelica officinalis*. Leaves in the form of an infusion, a decoction or as a chutney by grinding with black pepper were also used. These preparations were often prescribed for skin diseases, inflammation etc. and recently for diabetes. In the case of smallpox and measles, a decoction of neem leaves was given to quench thirst and to prevent dehydration. To make the surroundings of the patient aseptic and humid, the head was placed on a pillow made from neem leaves and small branches were hung all around and on the windows. In serious cases, instead of a decoction, leaf juice was prescribed and the whole body was smeared with a paste of neem leaves with or without neem oil. When pustules appeared, neem with liquorice (*Glycyrrhiza glabra*) root was made into a paste and applied.

For hemorrhoids, a paste of the leaves of neem and *Nerium indicum*, when applied, caused the shrinkage of inflammed tissues. For ulcers, neem paste acted the same way.

Neem was also used for snake bites, particularly the bite of a russel viper. It was said to destroy venom, but the effect appears to be due to delay in the clotting time of the blood.

In gynecological practice, the use of neem is quite popular in post-parturition disorders. It induces labor by uterine muscle contraction. As an anti-inflammatory and antiseptic agent, a decoction of neem is used for washing and douching the vagina. It induces the formation of milk, so was given for three days after delivery before meals. The same practice is followed by the dairy industry in some places in India even now. The effect appears to be hormonal because a neem branch is said to attract fish during the spawning period.

For oral hygiene and dental care, young twigs are cut into piece 15–20 cm long; one end of this twig is chewed to form a fibrous brush and the teeth are cleaned with it. During this process the juice of the twig gets mixed with the saliva in the mouth and the mixture clears the throat by irritating the mucus membrane and causing the expulsion of phlegm from the throat.

This practice is present even now, and it is a common sight to see fresh neem twigs being sold in the marketplaces in India (Fig. 15A), but this is proving disastrous for neem plantations, where even young plants are being destroyed for their twigs (Fig. 15B).

In the Siddha system (the traditional system practiced in some parts of south India), dried and cured neem leaves are used. The curing is said to make leaves more effective, more palatable and less toxic. These dry leaves are called *Vaipilla*, and a well-known Siddha medicated oil preparation is *Vaipilla tailam*.

Flower

Flowers as an item of diet were used in India and the other east Asian countries as a spinach and chutney. The flowers are said to expel worms, give relief from coughing and are considered good for the eyes. In cataracts, a suggested formulation is to make a very fine powder of equal parts of neem flower and potassium nitrate. The mixture is applied to the eyes.

Figure 15 *Azadirachta indica*, neem twigs. A, young twigs are being cut to have about 15 cm long and 0.5 cm thick pieces for chewing; B, young neem tree destroyed for the twig. (Photo by Gurcharn Singh)

Fruit and Seed

Unripe fruits are used in the same ways as leaves, but are considered useful for bleeding piles, worm infestation and urinary disorders. The ripe fruit is considered a blood purifier and anthelmintic and it makes the bowel soft.

For premature greying of hair, neem juice boiled several times in the juice of *Eclipta alba* is used as a nasal drop.

Oil

It was a common practice in some parts of south India to give a few drops of oil orally to infants regularly to keep them fit.

For skin diseases, the oil has always been considered a drug of choice and often used along with that of *karanj* (*Pongamia pinnata*) oil. It was applied on pustules, hard abscesses, obstinate types of wounds, leprotic lesions, ringworm, eczema and itch. It is particularly recommended for hair care problems such as psoriasis and dandruff, for killing lice and for giving relief in itching.

The oil destroys worms and may be useful in anal itching in infants caused by nematodes like the pin worm. It is a stimulative cerebral tonic and is used as a massage oil for rheumatism and joint pains.

Well-known Preparations of Neem

As given earlier, neem preparations have been prescribed for a very long time, but were not very much liked by patients because of their bitter taste. Recently, in some herbal patent preparations, neem extracts in the form of gelatin capsules are being sold, but the classical preparations are manufactured and supplied in the traditional way. The important ones are:

Nimba ghrit—general tonic in debility
Nimba asav—for fevers, particularly in malaria
Nimbadi anjan—antiseptic for eyes
Nimba arisht—tonic for building resistance in the body
Nimbadi kwath—decoction for fevers
Nimba Haridra—with turmeric and sugar, for throat problems
Nimbadi tail—oil for massage in dry eczema, leucoderma, and rheumatism
Panchnimba churn—for skin diseases like white patches, ringworm, etc.
Panchnimba gutika or *Panchamrit*—in leprosy and white patches
Panchnimba avleh—for skin diseases of different etiology, headache, diabetes and obesity
Panchtikata ghrit—in chronic skin diseases
Panchtikata ghrit guggal—for obstinate diseases, respiratory and heart problems

NEEM IN UNANI TIBB

In the Greco-Persian system of medicine (*Unani tibb*), which was patronized by Muslim rulers in the medieval era in the Indian subcontinent, the leaves and fruit were in the pharmacopoeia. As per this system, neem is cold 1°, dry 2°, a resolvent and blood purifier (Nadkarni, 1954). Neem leaves, called *"burgh-i-neem"*, are said to expel foul wind from the body and heal ulcers in the urinary passage; it is an emmenagogue and good for skin diseases.

Well-known Unani Tibb Preparations

Neem leaves, bark, seed and oil are incorporated in some of the Unani preparations (Said, 1970), which are as follows:

Arq Gaz—a distillate from all five parts of the neem tree, used for fevers due to inflammation of the spleen.
Arq Harabhara—a distillate from the seed coat, a tonic for the lungs
Arq Murakkab Musaffa khun—a distillate, has a cooling effect on the body, used for purification of blood in venereal disease
Hab Musaffi Khun—blood purifier for boils, itching etc.
Hab Narkachur—anti-inflammatory for children
Hab Bawasir Badi—for bloodless piles
Hab Siyah Chasham—for application inside the eyelid in conjuctivitis.
Majun Juzam—blood purifier for venereal diseases, leprosy etc.
Marham bawasir Jadid—ointment for external application on piles
Roghan Neem—neem oil for external application on sores and wounds, and for killing ectoparasites like lice
Zimad Bawasir—whole fruit powder, for application on hemorrhoids
Zimad Mohasa—for application on pimples, and other minor skin eruptions

NEEM IN HOMOEOPATHY

Neem under the name *Melia azadirachta* or *Melia Azadirachta* is well known in Homoeopathy for its bark, called *Margosa*. It was included in the *Pocket Manual of Homoeopathic Materia Medica* (Boericke, 1927). According to Ghose (*c.* 1930), the tincture of bark was introduced by Dr P.C. Majumdar, after proving by him and his pupil. A full report on this proving was published in the *Indian Homoeopathic Review*. Later on, two more provings were made and published. It was also included in *New, Old and Forgotten Remedies* (Anshutz, 1930), *Pathogenesis de Azadirachta indica* (Lamasson, 1968), *Indian Homoeopathic Pharmacopoeia* (1971) and *Materia Medica of New Homoeopathic Remedies* (Julian, 1979).

The *Indian Homoeopathic Pharmacopoeia* has described both the macroscopic and microscopic character of the bark. As per the pharmacopoeia, the bark is dark grey to greyish black externally, while the inner surface is pinkish brown and fibrous. In old bark there is a well-defined outer rind formed of alternating strips of cork layers and dead secondary bast. The cork cells have reddish brown contents. Phellogen is not very distinct and a secondary cortex is normally not present. The secondary bast is formed of sclerenchyma. Medullary rays are 2–5 seriate.

In the *Indian Homoeopathic Pharmacopoeia*, the drug has been introduced on the authority of *Drugs of Hindoostan* by Dr S.C. Ghose. The mother tincture (one liter) can be prepared by taking 125 gm of fresh bark (25 gm moisture and 100 gm solid matter), distilled water 375 ml, strong alcohol 635 ml. The fresh bark is pounded to pulp and macerated into alcohol. Potency 2x with dilute alcohol, 3x and higher with dispensing alcohol.

Findings of the Symptoms

A brief summary of the findings of the symptoms, as given by Ghose (c. 1930) are:

Mind: oppressed, forgetfulness, dull and loss of memory
Head: giddiness, headache
Eye: burning, dull and heavy, painful, red
Ear: buzzing in the ear
Face: flushing
Stomach: thirst, appetite very acute
Abdomen: great uneasiness with flatulence
Stool: insufficient bowels, constipation, stool hard
Genitourinary organs: great excitement of sexual organs
Respiratory organs: very troublesome cough, sputa white
Pulse: quick and hard
Extremities: numbness of the limbs, burning of hands and soles
Sleep and dreaming: sleeplessness, dreams of quarrels
Fever: glowing heat and burning, copious sweat, itching

Julian (1979) has also given the symptomatology, according to which neem acts in active, depressed and forgetfulness states, right-sided headache, insomnia, thirst, constipation, fevers, tendency to miscarriage. The clinical diagnoses are senile dementia, amnesia, hypochondria, acute articular rheumatism, acroparesthesia, typhoid fever, recurring of miscarriage and metritis of the cervix.

REFERENCES

Anshutz, C. (1930) *New, Old and Forgotten Remedies*. vide Ghose, S.T. (c. 1930), *Drugs of Hindoostan*, Hahneman Publishing Co., Calcutta.
Boericke, W. (1927) *Pocket Manual of Homoeopathic Materia Medica*. Boericke and Runyon, New York.
Clarke, J.H. (1900) *Dictionary of Practical Materia Medica* (reprint edition). B Jain Publishers, Delhi, India.
Dey, K.L. (1896) *The Indigenous Drugs of India* (reprint edition). International Book Distributors, Dehradun, India.
Ghose, S.T. (c. 1930) *Drugs of Hindoostan*. Hahneman Publishing Co., Calcutta, India.
Indian Homoeopathic Pharmacopoeia (1971) Ministry of Health, Government of India, New Delhi.

Julian, O.A. (1979) *Materia Medica of New Homoeopathic Remedies*. Beaconsfield Publishers Ltd, Bucks, UK.

Lamasson, F. (1968) Pathogenesis de Azadirachta indica. *Annales Homoeopathic Francaises*, **10**, 3/851–10/858.

Majumdar, P.C., *Indian Homoeopathic Review*, **3**, *vide* Ghose, S.T. (c. 1930) *Drugs of Hindoostan*. Hahneman Publishing Co., Calcutta, India.

Nadkarni, K.M. (1954) *Indian Materia Medica*. Vol. I. Popular Prakashan, Bombay.

Said, H.M. (1970) *Hamdard Pharmacopoeia of Eastern Medicine*. Hamdard Foundation, Karachi, Pakistan.

10. THERAPEUTIC INDICATIONS AND PHARMACOLOGICAL STUDIES

The effects of bitters on human systems are well known. Much of the earlier therapeutic efficacy claimed for neem may be due to these bitters. It was included in the Indian Pharmacopoeial List as a bitter preparation which acted as a sialogogue and a stimulant of various gastric secretions due to reflex action of the body against the bitter taste, which in turn increased the appetite.

In *Potter's New Cyclopaedia of Botanical Drugs and Preparations* (Wren, 1907), neem bark is mentioned as an anthelmintic, cathartic and emetic, used for children in the southern states of America. It was prescribed in the form of a decoction, followed by purgation by castor oil. These and other studies, as given in the section on Ayurveda, Unani-Tibb and Homeopathy and in folklore medicine, led to research on the pharmacological activity of neem preparations on the various systems of the human body, and human pathogens. The details of these studies are given here. Use in veterinary practice has been given separately.

ANTHELMINTIC ACTIVITY

Keeping in view the reputation of neem as an anthelmintic, Cauis and Mhaskar in 1923 administered leaf juice to patients in a dose of 4 drams (14.25 gm) for expelling intestinal worms, followed by purgation, but the treatment was ineffective (Nadkarni, 1954); on the other hand, an Ayurvedic compound (Tyagi *et al.*, 1977) containing neem and long pepper (*Piper longum*) gave good results in ankylostomiasis. Singh *et al.* (1980) carried out further clinical evaluation of the anthelmintic activity of neem.

Recently a large number of studies in plants have confirmed the nematodocidal effect of neem and it should be effective against nematodes in the human body, like round worm (*Ascaris lumbricoides*), threadworm (*A. vermicularis*) and pin worm (*A. oxyrus*). The activity of neem against pin worm in particular is of interest because the worm causes irritation and itching in the anal region, mostly in infants and children, in many parts of the world. For symptomatic relief, sometimes mustard oil is applied in the anal region, which may have an antifeedant property. Neem oil appears to be a better substitute for mustard oil for this purpose.

Neem was tried against the nematode *Ascaridia galli* (Shilaskar and Parashar, 1989). An Ayurvedic antifilarial compound with neem as a major ingredient has given good results against *Acanthocheilonema vitae*, *in vitro* (Comley *et al.*, 1990).

ANTIBACTERIAL

Neem preparations have been used as a disinfectant since ancient times. With the discovery of antibiotics from the lower organisms, there was a move to look into higher plants for this type of compound. Chopra *et al.* (1952) tested essential oil from flowers against a number of human and plant pathogens. Most of the organisms, except *Bacillus mycoides*, showed sensitivity to this essential oil. Joshi and Nagar (1952), Bhat (1957) and Ramaniyalu (1986) worked on the antibacterial activity of neem oil. Patel and Trivedi (1962) also carried out these studies, while Ramaswamy and Sirsi (1967) worked on the antitubercular effect of neem. Rao and Satyanaryana (1977), Singh and Sastry (1981), Vinayagamoorthy (1982) and Kher *et al.* (1984) when tested, failed to demonstrate any antibacterial activity from any fraction of neem. Bhatt and Koro (1984) worked on the antibacterial effect of neem on wound sepsis. Thaker and Anjaria (1986), Jain *et al.* (1987), Rao *et al.* (1986), Vashi and Patel (1988) and Grand (1989) studied antibacterial activity. Siddiqui *et al.* (1992) isolated an antibacterial compound, mahmoodin and naheedin. Souza *et al.* (1993) found antimicrobial activity in an aqueous extract.

Prasad *et al.* (1993) reviewed the earlier literature and concluded that the extracts can significantly inhibit pathogenic microorganisms, with the most potent effect against *Salmonella typhi*, *S. paratyphi* and *Bacillus subtilis*, *in vitro*. The leaf and bark extract showed activity against *Staphylococcus epidermidis*, *S. aureus*, *Bacillus* spp. and *Proteus* spp. The activity against *Escherichia coli* was moderate (Ahmad *et al.*, 1995). Garg *et al.* (1995), while looking for the antimicrobial spectrum for a neem-based cream developed by them, enumerated the following bacteria, on which the cream had a bactricidal effect: *Bacillus subtilis*, *B. anthracis*, *Corynebacterium* spp., *Escherchia coli*, *Micrococcus* spp., *Proteus vulgaris*, *Staphylococcus aureus*, *S. citrus*, *S. epidermidis*, *S. lactitis*, *Salmonella paratyphi* A and B, and *S. typhi*.

ANTIDIABETIC

Neem is one of the indigenous Indian drugs used in the treatment of diabetes (Mukherji, 1957). Shukla *et al.* (1973) observed that oral administration of 5 g of an aqueous extract or equivalent amount of dried leaves in capsules enabled patients to reduce their dosage of insulin up to 30–50 percent. Luscombe and Taha (1974) tried a 10 percent aqueous extract of dry young tender leaves for hypoglycemic effect in rabbits and a marked fall in blood glucose concentration was observed. The same effect was seen in fasting rats and guinea pigs. It appeared that antidiabetic activity was dependent on the functioning pancreatic beta cells, since neem leaves did not produce hypoglycemia in totally pancreatectomized rats or animals made severely diabetic by alloxan. Neem extract appeared to behave the same way as sulphonylurease. Murty *et al.* (1978) on the other hand, when they administered intravenous aqueous leaf extract to dogs, observed a significant decrease in blood glucose level in both normoglycemic and adrenaline induced hyperglycemic animals (Anonymous, 1979).

The effect of oral doses of an aqueous leaf extract, seed oil and nimbidin (the bitter principles from neem) was investigated by Pillai and Santhakumari (1981a, 1981b). The aqueous extract was found to be inactive but seed oil and nimbidin exerted significant activity. Sharma *et al.* (1983) also studied blood sugar level in hyperglycemic and diabetic animals. Management of diabetes mellitus by neem was discussed by Upadhyay (1984). Chakraborty and Poddar (1984) observed that a water extract of neem leaves exhibited a blood sugar lowering effect and this was statistically significant as compared to hyperglycemic activity induced by streptozotocin (STZ). Obaseki et al. (1985) suggested that an aqueous extract of neem may have a therapeutic action via interaction with membrane. There was a reduction in serum acid phosphate activity and an increase of 5' nucleotidase. Neem oil was also found to have a significant effect on lowering the blood glucose level of normoglycemic and hyperglycemic rats by Dixit et al. (1986). Neem as a synergistic agent to antidiabetic drugs was suggested by Bhargava (1986, 1987), Bhargava et al. (1986). Chakraborty et al. (1989) evaluated the activity of neem leaf extract for hypoglycemic activity in rats. El-Hawary and Kholief (1990) carried out studies on the biochemical effect of neem leaf. Dixit et al. (1992) found that neem oil reduced blood glucose levels in normal animals as well as in alloxan-induced diabetic rats.

Chattopadhyay et al. (1993, 1993a) observed that neem leaf extract failed to improve muscle glycogen content and did not influence the effect of exogenous insulin. It did not affect the hepatic glycogen content of normal rats but reduced it in the case of glucose-fed hypoglycemic rats.

Literature on the antidiabetic activity of neem has been reviewed by Handa et al. (1989), Nat et al. (1991) and Prasad et al. (1993). According to Nat et al. (1991), the antidiabetic mechanism of action may be due to the release of indigenous insulin in a way similar to that of sulphonyl urea. Prasad et al. (1993) put forth the view that neem only produced moderate-grade hypoglycemia. Extracts induced a significant fall in glycemic levels in a model of glucose-induced hyperglycemia in rats, had less effect on moderate alloxan diabetic rats but was ineffective in severe alloxan diabetic rats, probably due to the complete destruction of β cells by alloxan.

ANTIFERTILITY ACTIVITY

During pharmacological studies, Murthy and Sirsi (1958a) observed estrogenic activity in neem oil and its derivatives. Two derivatives of neem oil, sodium nimbinate and sodium nimbidinate, were found to possess a spermicidal effect *in vitro* in human beings and in rats (Sharma and Saksena, 1959, 1959a). Deshpande et al. (1980) gave a preliminary report on the male antifertility effect of neem in mice. Oral administration of an extract of crushed neem leaves for one month produced reversible antifertility activity in animals without inhibition of spermatogenesis. Chaudhry and Haq (1980) also observed this activity. Sinha et al. (1984) reported antiflagellate activity of the sperm by neem oil. The sperm became immobile and were not able to fertilize the ovum. Undiluted neem oil was found to possess a strong spermicidal activity both *in vitro* and *in vivo* in both rhesus monkey and man. Once the sperm touched the margin of oil it got completely entangled, leading to further sluggishness culminating in immobility within thirty seconds. All the animals that mated after the application of neem oil showed no sign of implantation. The application of oil did not produce any irritation or side effect. In females, the application of oil increased uterine contraction and there was suppression of ovulation, with an antiimplantation effect (Sinha et al., 1984a).

Reddy et al. (1984a) observed that neem oil affected the estrous cycle in albino rats; it had an antiovulatory and antifertility effect (Reddy et al., 1984b). Salunkhe and Adsale

(1985) considered neem oil as one of the agents for population control. Lal *et al.* (1986) confirmed the antifertility effect of neem oil in female albino rats by intravaginal and oral routes. To study the mechanism of the antifertility effect, a freshly prepared aqueous extract of crushed neem leaves was fed to rats by Mateenuddin *et al.* (1986). There was no significant loss or gain of weight in rats due to this extract. The absence of a open vagina, epithelial cells in the vaginal smear and incomplete development of the uterus indicated that there is no oestrogenic activity in neem leaf extract.

Subcutaneous administration of 0.2 to 0.3 ml/kg of neem seed oil for 1–5 days postcoitum terminated pregnancy in rats (Tewari *et al.*, 1986). These authors, on the basis of further biochemical and histological studies of the reproductive organs in cyclic and ovaricotomosed rats, confirmed that the postcoital contraceptive effect of neem oil was non-hormonal. It appeared that the active compounds of neem exerted the antifertility effect by absorption by the vaginal tissue and later on by circulation in the blood (Riar *et al.*, 1988).

The non-hormonal nature of neem oil was further confirmed by Prakash (1986) and Prakash *et al.* (1987, 1988). Sharma *et al.* (1987) also observed antiandrogenic properties of neem seed oil in both males and females. The testes of rats treated with neem oil showed that the oil caused a reduction in their weight, arrested spermatogenesis and brought about severe degenerative changes in the cauda epididymus with a decline in protein, acid phosphate, etc. The effect appeared to be antiandrogenic, without any estrogenic, anti-estrogenic or progestinol activity. It did not interfere with the action of progesterone. In females it acted by causing damage to the uterine histological structure, especially breakage of the endometrium (Prakash *et al.*, 1988).

The antifertility effect of the oil appeared to be due to:

1. A spermicidal or ovicidal effect so that the zygote is not formed.
2. If the zygote is formed, neem oil is blastocidal; a dead blastocyte cannot get implanted in the uterus.
3. Even if the blastocyte is formed, it is not able to establish itself because of the denatured endometrial linings of the uterus.

Further studies were carried out by Prakash *et al.* (1991, 1991a) to study the effects of the ethanolic extract of the seed on female rats. Bardhan (1991) tried oil in rhesus monkeys and found a good spermicidal activity after pre- and postcoital application of the oil intravaginally, while Garg *et al.* (1992) carried out research on spermatogenesis by solvent-extracted, water-washed neem seed cake.

Upadhyaya *et al.* (1990) put forth a concept of immunocontraception, according to which a single intrauterine treatment led to a long term antifertility effect in female rats. The effect lasted for a variable period of 107–180 days and was reversible in nature.

Nat *et al.* (1991), after a review of the literature, concluded that the antifertility activity reported both for neem seed oil and leaves pointed at limonoids, which occur in both plant parts and are the active constituent. The existing relationship between the structure of some steroids and triterpenoid of neem suggests hormonal regulation of fertility control, but the non-hormonal mode may also be playing a part.

Shaikh et al. (1993) also confirmed the antispermatic activity of neem but some other studies have shown that use of neem oil as a contraceptive is not safe but has serious side effects. Sampathraj et al. (1993) confirmed marked structural alterations in the testes by administration of neem oil. Spermatogenesis was drastically impaired with an adverse effect on the testicular function and also as the function and integrity of the epididymus. Parshad et al. (1994) studied the effect of neem by oral administration. It resulted in a significant decrease in the total testosterone, bilirubin, and K^+ in serum. There was no cytotoxicity.

Keeping the above side effects in view, Garg et al. (1993), at the National Institute of Immunology in India, looked for the safe fractions from neem oil for contraception. Garg et al. (1994) investigated the hexane extract of neem seed. Paranjape and Paranjape (1993) tried neem oil suppositories for contraception at the pre-coitus stage. The preparation was not irritating to the vaginal mucosa or the male genital organ, and was very safe.

Kasturi et al. (1995) studied histological and biochemical changes in the caput and cauda didymus of albino rats treated with dry leaf powder in various doses for 24 days. The effect was dose dependent. The serum testosterone concentration in the animals treated with high doses decreased significantly, suggesting a possible antiandrogenic property. Neem seed oil was applied by 77 human fertile females before coitus. It covered 6341 menstrual cycles. There was no contraception in any case (Tyagi and Sahrawat, 1995). The safe fraction of neem oil developed at the National Institute of Immunology in India for contraception was tried by Mukherji et al. (1995, 1995a) by oral administration. All the embryos without exception were resolved. The treatment was well tolerated with no mortality or morbidity. Animals regained fertility in subsequent cycles. Talwar et al. (1995) in the same institute developed a polyherbal cream and pessary, containing extract of neem, quinine hydrochloride, and soap nut extract (*Sapindus mukrosii*). The formulation showed a good spermicidal action *in vitro* and a high contraceptive efficacy in rabbits and monkeys. An acute and sub-acute toxicology study of this cream was carried out and no change in any of the hematological and biochemical parameters compared to pre-treatment values has been observed. Autopsy of animals revealed no gross changes in the tissues (Garg et al., personal communication). This cream is now being subjected to multi-centric clinical trials.

ANTIFUNGAL

In the case of infectious diseases, it is a common practice among some people to use neem with other ingredients as a fumigant. To find out the significance of this, Murthy and Sirsi (1957), Naryana (1965), David (1965) and Jain and Pathak (1970) studied antifungal activity. Upadhyay and Arora (1975–76) studied the sporostatic nature of neem smoke and its possible ecological influence on the air fungal flora. Some people hang leaves around the patients. The curative effect may probably be due to essential oils. To confirm this, Satyanaryana and Rao (1977) and Thind and Dahiya (1977) tried the essential oil obtained from neem leaves against keratinophilic fungi. Singh and Sharma (1978) and Chary et al. (1984) also studied antifungal activity.

An account of raw materials from neem, which can be used against fungi pathogenic to man, has been given by Khan and Wassilew (1987). Dube and Tripathi (1987) confirmed that the bark and leaf extracts inhibited spore formation and were toxic against *Epidermopyton flocossum*, *Microsporum canis*, and *Trichophyton mentagrophytes*, but Singh *et al.* (1987) failed to demonstrate any antifungal activity of two fractions of leaf, extracted with acetone, against a variety of human pathogens. Khan *et al.* (1988) also observed that neither dried neem material nor medicinal preparations containing neem or its oil had any effect on fungal growth. Petroleum ether extract showed some activity, which may be due to quercetin—a flavonoid.

Khan *et al.* (1991) also studied the effect of petroleum ether extract of neem on the fungi pathogenic to humans. Iyer and Williamson (1991) concluded that neem extract inhibited the protease activity of *Trichophyton* spp.

Prasad *et al.* (1993), on the basis of earlier reports, concluded that neem extract exerted varying degrees of fungal toxicity which may be due to volatile sulfurous compounds or the limonoid gedunin. Fabry *et al.* (1996) also studied fungistatic and fungicidal activity.

ANTIINFLAMMATORY AND ANTIPYRETIC

In Ayurveda, Nimbadi Kashyam (a decoction of neem with other herbs) is prescribed for various inflammatory conditions. Other neem preparations are also used as a poultice for external application in gout, rheumatism, arthritic pains, etc. Neem oil was found to be an antipyretic and effective against rheumatism by Murthy and Sirsi (1958). David (1969) studied the antipyretic activity of neem oil, while Bhargava *et al.* (1970) worked on its antiinflammatory activity. Lorenz (1976) tried neem bark extract in inflammatory stomatis and David (1978) the effect of neem oil and its constituents on cotton pellet inflammation. Shankarnaryana (1978) used two crystalline bitter principles, nimbin and nimbinin, obtained from oil and found that they were comparable to cortisone in their action. Crude neem oil, sodium nimbidinate and nimbidol also showed considerable antiinflammatory activity. Pillai *et al.* (1980) observed that nimbdin 100 mg/kg showed a significant analgesic and antipyretic effect in mice and rats which was comparable to acetylsalicylic acid and pethidine hydrochloride. Further studies by Pillai and Santhakumari (1981) indicated that nimbidin reduces acute paw oedema significantly in rats and also suppressed formalin-induced arthritis of the ankle joint and fluid exudation in croton oil-induced granuloma in rats. The authors were of the view that bitter from the neem may be suppressing oedema formation by reduction in vascular permeability and was not mediated through adrenalins.

An antiinflammatory (rat paw oedema) and fairly good antipyretic effect (pyrogen-induced hyperpyrexia) with 75 percent methanolic leaf and bark extract in rabbits was observed by Okpanyi and Ezeukwu (1981). Two patents were granted to Terumo Corporation of Japan for anti-inflammatory polysaccharides from the bark in 1983 and 1985. Fujiwara *et al.* (1982, 1984) also reported inhibition of caragenin-induced oedema after oral administration of polysaccharides isolated from bark. Khatak *et al.* (1985) obtained an antipyretic effect from various fractions of 90 percent ethanolic leaf and twig extract. Tidjani *et al.* (1989) observed antiinflammatory activity from chloroform extract when administered orally and also applied topically. Tandon et al. (1990) reported that the ethersoluble fraction of ethanol extract of leaves showed a good analgesic activity in pain but did not have any anti-inflammatory effect. Handa et al. (1992) and Vohora and Dandiya (1992) have reviewed the literature on antiinflammatory activity. Chattopadhyay et al. (1994) have also discussed it.

According to Bray et al. (1990), relief due to neem preparations in inflammation, pain, and fevers may be due to interference with hormonal regulation, interaction with receptors and alternation of membrane permeability and integrity. Nat et al. (1987) postulated the view that the inhibition of calcium activation and chemiluminescence *in vitro* correlates with the antiinflammatory and antirheumatoid effects. While continuing this study, Nat et al. (1991a) identified the compounds responsible for the inhibition of the chemiluminescence product by activating human polymorphonuclear leukocytes. These compounds are gallic acid (+) gallocatechins, (−) epicatechins, and (as a 2:1 mixture) (+) catechin and epigallocatechin. Commercial samples of gallic acid, (+) catechin and (−) epicatechin showed the same effect. Kroes et al. (1992) confirmed the antiinflammatory activity of gallic acid.

Khanna et al. (1995) studied the mechanism involved in the antinociceptive action of neem. The results suggested that both central and peripheral mechanisms and complex pathways of both opioid and non-opioid are involved.

ANTIPROTOZOAL

Neem has been tried against the malarial parasite and *Trypanosoma*.

Malarial Parasite

Neem preparations, particularly bark, have been esteemed for malaria for a very long time. Studies were conducted to see if it could be substituted for cinchona bark. A compound of margosic acid and its sodium salt was prepared from the oil and on the basis of clinical trials was found to have activity against protozoa, which was one-tenth that of quinine sulphate (Nadkarni, 1954). Murthy and Sirsi (1958b) experimented on avian malaria. In recent times, neem has been investigated in detail for its antimalarial activity but the results are conflicting. Tella (1976) found oral doses of neem preparations quite ineffective but Ekanem (1976) demonstrated the schizonticidal activity of these against *Plasmodium berghei berghei*. Nimboloid, a terpenoid lactone, inhibited *P. falciparum* in culture with a moderate potency (Rochanakij et al., 1985). Obaseki et al. (1985) studied the effect of aqueous extract in rats and suggested that it may elicit its therapeutic action via an interaction with membrane. Obih and Makinde (1985) administered neem extract subcutaneously into mice after inoculating with a chloroquin-resistant strain of *P. berghei berghei*. There was appreciable suppression of parasitemia. Bray (1985) studied the antimalarial activity of some limonoids.

Abatan and Makinde (1986) screened solvent fractions obtained from leaves for antimalarial action, using *P. berghei berghei* as a parasite in mice. Statistically significant suppression was seen after four days of oral dosing. Obaseki and Jedge-Fadunsin (1986) administered aqueous extract of neem to albino mice. It proved effective against acute infection by *P. yoelli nigeriensis*. The flavonoids and limonoids, gedunin and quercetin isolated from neem were found to be highly active against the malarial parasite (Khalid *et al.*, 1986; Khalid and Duddeck, 1986).

When Iwu *et al.* (1986) studied levels of some enzymes in rat liver microsome after administering the water-soluble part of methanolic extract, a significant inhibition of nictinamide adenine dinucleotide phosphate (NADPH) cytochrome C (P/450) reductase activity was observed. It also stimulated the oxidation of hemoglobin and glutathione. The leaf extract and seed extract of neem were found active against chloroquin-sensitive and resistant strains. The extracts were non-toxic.

The potential of natural plant products for treating major protozoal diseases, with reference to neem and other trees, was reviewed by Phillipson and O'Neill (1989). These limonoids were tested for their *in vitro* antimalarial resistant K1 strain of *P. falciparum*. Out of all these compounds, gedunin was three times more active than chloroquin. Vasanth *et al.* (1990) studied neem as one of the antimalarial plants, while Udeinya (1993) looked into the antimalarial activity of neem leaves. Jones *et al.* (1994) observed that the sexual development of the malarial parasite is inhibited *in vitro* by neem extract, azadirachtin and its analogs. Iwalewa *et al.* (1995) studied cardiogenic glycosides and sterols from neem on *P. yoelii nigerensis* for antimalarial activity. These compounds exhibited high prophylactic, moderate suppressive and very minimal curative properties. Cardiac glucosides were more effective than sterols.

The antimalarial activity of neem appears to be due to other mechanisms exerted on the body, in addition to that on the malarial parasite. According to Okpako (1977), neem extracts have been shown to break the periodic sequence of recurring fevers because of its prostaglandin synthetase inhibitory action. The body of the patient mobilizes the immune system to suppress the parasite. Badam *et al.* (1987) and Benoit *et al.* (1996) found an antimalarial effect of both leaf and seed *in vitro*. Bray *et al.* (1985, 1990) also considered the beneficial effect in malaria due to anti-inflammatory and immunomodulater activity. Nat *et al.* (1987), after their study, concluded that it may be due to immunomodulater activity in bark and an antipyretic effect in other parts of the tree (Nat *et al.*, 1991).

Trypanosoma

Neem was tested against Chagas' disease, which spreads in endemic form in some parts of Latin America. Garcia (1989) studied the effect of azadirachtin on the development of *Trypanosoma cruzi*, a flagellate protozoa found in the blood, and a causative agent of Chagas' disease and its vector *Rhodnius prolixus*. It affected the hormonal balance of host (Azambuja, 1991) and caused growth inhibition of the parasite. Nok *et al.* (1993) confirmed the trypanocidal effect of neem leaf extract.

Two other diseases, tsetse fly and sleeping sickness, are caused by *T. bruci* and *T. gambiensis* respectively. Neem should be active against these parasites also.

ANTITUMOUR

Some earlier reports (Chatterjee, 1961; Anonymous, 1980) indicated the anticancerous activity of neem. Mitotic inhibition activity by the leaf extract was observed by Yadav and Rathore (1976). *In vitro* activity against sarcoma 180 ascites tumor by intra-peritoneal administration of polysaccharides from the bark was reported by Fujiwara *et al.* (1982, 1984), while Pettit *et al.* (1983) studied the effect of limonoids against the P-388 lymphocytic leukemia system. Out of the limonoids investigated, $1\beta,2\beta$ diepoxyazadirdrone and 7-acetylneo-trichlenone appear to be active. Antitumor polysaccharide (code-name N9C1) which, when administered to mice, markedly inhibited the growth of sarcoma 180 was patented by Terumo Corporation in 1983 and 1985. In the first patent, organic solvents were used to get the extract but in the second, neem bark was treated with water at a temperature between 0° and 40°C. The residue was subjected to a purification process by alcohol precipitation.

In their review article on neem, Prasad *et al.* (1993) mentioned another patent for antineoplastic action. Its usefulness for mouth cancer and its cytocidal effect on malignant cells have been documented. In another report, it was mentioned that azadirachtin had an immediate effect on the rate of cell proliferation, including its intensity. This was due to rapid inhibition of RNA synthesis along with inhibition of the synthesis of protein and DNA. Takeya *et al.* (1996) and Cohen *et al.* (1996) have reported the cytotoxic effect of limonoids.

ANTIULCEROGENIC

Nimbidin, a mixture of neem bitters, was studied by Pillai (1978) for antigastric activity and was found to be effective. Pharmacological studies showed that nimbidin blocked the stimulatory effect of acetylcholine and also inhibited the stimulation produced by histamin. It partially blocked the action of nicotine on the arterial blood pressure of dogs under the effect of anesthesia. All these findings suggested the antigastric activity of nimbidin. Nimbidin, when given orally, significantly prevented Shay ulceration. These observations were confirmed by Pillai and Santhakumari (1984), and further studies showed the significant protective effect of nimbidin in a dose of 20–40 mg/kg. It afforded a remarkable protection in duodenal lesions. Oral and intra-peritoneal administration of nimbidin did not produce toxic manifestation or foetal abnormality (Pillai and Santhakumari, 1984a).

Koley *et al.* (1994) studied the chronic ulcerogenic effect of alcohol extract of neem leaf on the gastric mucosa of rats. The lack of irritation and promising antiinflammatory effect suggested its use as an antiulcerogenic or as an antipeptic ulcer drug. Garg *et al.* (1993) and Devdas *et al.* (1995) also observed the gastric antiulcer effects of neem leaves.

Nimbatiktam, a bitter principle of neem containing terpenic ester (nimbdin), was administered at the rate of 30 mg daily. It showed a significant ulcer-healing effect. Endoscopy healing was 87 percent for duodenal ulcer and 100 percent for gastric ulcer, with significant improvement from the third week of medication. No side effects were observed (Pillai, 1995).

ANTIVIRAL

Rao *et al.* (1969) noted that ten percent water extract of tender leaves possessed antiviral activity against *vaccinia* and *variola* viruses. Rai and Sethi (1972) found that an aqueous extract of neem leaves did not kill the virus directly but inhibited the multiplication of *Vaccinia* and fowl pox virus. They suspected that this effect may be due to limonoids or flavonoids. Babbar *et al.* (1982) also found antiviral activity in neem leaves. Reddy and Sethi (1984) studied, *in vivo*, the antiviral effect against *Vaccinia* virus while Wagh (1988) used it in clinical trials in the case of viral hepatitis.

Out of 15 viruses studied by Gogate and Marathe (1989), only three, *Chikungunya* (634029), *measle* (ED/3) and *Vaccinia* virus, were inactivated by neem; others were affected to various degrees. Praneem polyherbal cream, which has been developed by the National Institute of Immunology in India and in which neem extract is the main ingredient used, has been found to be effective in the treatment of human papillom virus-16, which infects the female genital tract (Talwar *et al.*, 1995).

The above antiviral effect of neem in some cases justifies the use of neem leaves, particularly in smallpox in the Indian subcontinent.

CARDIOVASCULAR EFFECTS

Thompson and Anderson (1978) studied a crude extract of neem on the cardiovascular system. The effects included profound hypertension and a minimal negative chronotropic effect, which increased at higher doses. The rise in arterial blood pressure with low doses of the extract supports the suggestion that it may have a biphasic effect on arterial blood pressure. The vasodilation remained somewhat persistent. Ilesanmi *et al.* (1988) also found cardioactivity in rats with neem leaves.

DIURETIC EFFECT

Bhide *et al.* (1958) found diuretic activity in sodium nimbidinate. It was seen by Shah *et al.* (1958), who carried out clinical trials with it in nine cases of congestive heart failure, that patients obtained relief due to the diuretic action of neem preparations. There was no local discomfort, no toxic effect and the diuretic effect continued for a few days. Luscombe and Taha (1974) also confirmed that neem leaves have a mild diuretic property as observed in water-loaded rats. Singh *et al.* (1987), using oral administration of neem leaf, observed a significant reduction in blood pressure as well as heart rate but no diuretic activity.

EFFECT ON CENTRAL NERVOUS SYSTEM

Debelmas and Hache (1976) tested the aqueous extract on the central nervous system (CNS). No significant effect was observed. No anticonvulsant, anticholinergic, analgesic, or sedative effect could be demonstrated, but Pillai *et al.* (1980) showed an antistress property. Pillai and Santhakumari (1984a) observed a mild suppressive effect on the CNS functions of mice with nimbidin (the bitter principle of neem). Singh *et al.* (1987) noted that acetone extract of leaf had a variety of effects on mice which were attributed to CNS depression and to an effect on the autonomic nervous system. Dandiya (1990) also considered neem as one of the herbs acting on CNS.

In a continuation of their early study, Singh et al. (1987) worked on two fractions of acetone extract of neem leaves and observed a CNS-depressant activity. Jaiswal et al. (1994) mentioned an anti-anxiety effect (anxiolytic) of leaf extract in rats.

While summarizing the earlier findings, Nat et al. (1991) put forth the view that the active constituents from the leaf extract showing CNS activity may be limonoids, because most of these compounds are sufficiently lipophilic to cross the blood brain barrier. The acetone extract displayed a much stronger activity than an aqueous leaf extract. According to Nat et al. (1991), considering the antihistamine properties of nimbidin, interference with CNS neurotransmitters may be possible.

HEPATOPROTECTIVE EFFECT

Ayurvedic polyherbal preparation "Nimbatwagadi kashyam", containing neem bark and other herbs, is specific for jaundice. Oomacham et al. (1990) reported neem as one of the plants used for the treatment of jaundice. To find the hepatoprotective action of neem, Chattopadhyay et al. (1992) studied serum levels of glutamate oxaloacetate transaminase (GOT), glutamate pyruvate transaminase (GPT), acid phosphate and alkaline phosphate to elucidate liver damage. The levels of these enzymes were elevated after 24 hours of paracetamol treatment, but rats fed with neem extract prior to addition of paracetamol to their diet showed a much lower serum level. The hepatoprotective activity may be due to 6 flavanol-o-glycosides.

IMMUNOMODULATER PROPERTY

The antistress property of nimbidin, a compound isolated from neem oil, was noted by Pillai (1978). In an initial screening of Sri Lankan plants *in vitro*, neem bark was found to have an anticomplementary effect (Nat et al., 1987), which exerted a dose-dependent effect on immunomodulater activity. A dose-dependent decrease in the chemiluminescence of polymorphonuclear leukocyte was observed along with a dose-dependent increase in the production of the migration inhibition factor by lymphocyte. When the immunomodulater property of aqueous extracts on the human humoral and cellular defense mechanism was investigated, the extracts decreased both the classical and alternative C pathways. Labadie et al. (1989) reported the immunomodulater activity of a compound isolated from neem bark on the basis of data collected from various sources. Nat et al. (1989) characterized the two-polymer anticomplement compound as peptide glycans. The carbohydrate part consisted of glucose and arabinose. The protein content was 5.5 to 9.8 percent. Compounds responsible for the inhibition of chemiluminescence production by activated human polymorphonuclear leukocytes were isolated and identified from the crude aqueous bark extract. The compounds were gallic acid (+) gallocatechin, (−) epicatechin, (+) epicatechin and epigallocatechin (Nat et al., 1991a).

An Ayurvedic preparation, Nimbarishta—a potion prepared by fermenting *Woodfordia fruticosa* flowers, neem bark and the other herbs—was studied for its immunological property by Kroes et al. (1990). These authors modified the preparation process of Nimbarishta to see the effect on complement activity and functioning granulocytes. It was seen that a long boiling time considerably reduced, whereas the addition of flowers of *Woodfordia fruticosa* considerably increased, its immunomodulatory activity (Kroes et al., 1993). Gopal Raj (1993) mentioned that neem boosts the body's defense. Sen et al. (1993) observed the adaptogenic effect. Upadhyaya et al. (1993) studied extract of leaf bark and seed. The result indicated that neem was an immunostimulant; it selectively activated the TH-1 component of the lymphocyte population to elicit an enhanced cell-mediated immune response. *In vitro* HIV antieffects were observed. The addition of neem leaf extract significantly reduced the secretion of P-24 viral protein into culture supernatant. Bark extract was most effective. It also induced *in vitro* production of 1L-1 interferon. Garg et al. (1993) felt that lipid terpenoid association was responsible for immunomodulation activity. Sen et al. (1993) evaluated neem on some biochemical parameters in normal and stressed rats and found changes comparable to the antistress agent dopamine. Upadhyaya and Dhawan (1994) observed that an aqueous extract of leaf, bark and seed enhanced the phagocytic activity of macrophages.

USE IN SKIN DISEASES

On the basis of the earlier studies, Nadkarni (1954) gave an account of various neem preparations used for skin diseases. One of these was sodium and potassium margosate, derived from margosic acid, isolated from neem oil. It was found to be a disinfecting agent for skin infections and was used for dressing in place of carbolic acid. Its major use was for external application in tetanus, leprosy, urtica, eczema, eryspelas, scrofula, ringworm, scabies, etc. but intramuscular administration was more effective. For leprosy, it was used alone or mixed with *chawlmogra* oil (*Hydnocarpus kurzii*). Sodium margosate was administered subcutaneously and intramuscularly in the primary, secondary and tertiary stages of syphilis but it did not compare well with heavy metal preparations available at that time (Nadkarni, 1954).

Nimbidin, the bitter principle from neem, was found effective in furunculosis and arsenical dermatitis (Singhal and Mudgal, 1983).

An oil called Eth Ennei in the Sidha system of medicine, prepared by boiling nux vomica (*Strychnos nux vomica*) seed in neem oil, gave good results in eczema (Masilamani et al., 1993). An extract of dry leaves prepared with 70 percent alcohol was dissolved in propylene glycol. This preparation was applied in chronic skin diseases such as eczema, both acute weeping and acute chronic, ringworm and scabies. Most of the cases were of an obstinate type, treated earlier with well-known preparations like salicylic acid, benzoyl benzoate, sulfur, etc. Singh et al. (1979) got very encouraging results from this preparation.

A neem-containing formulation for hair was patented by Latiff (1979). The effect of nimbidin on psoriasis was studied by Rajasekharan et al. (1980). Shrivastava and Singh (1982) carried out research on the activity of neem against dermatophytes. Barde and Singh (1983) studied the activity of neem preparations against *Scytalidium anamorph* and *Hendersonula toruloidea* causing skin and nail diseases in man. Banerji and Nigam (1983) worked on the antiproteolytic activity of triterpenoids. Mahajan (1986) tested 18 fungi, reported to produce keratomycosis in man, with water and alcohol extracts of neem and found these effective. Nair *et al.* (1987) clinically evaluated an Ayurvedic preparation in vitiligo. Dube and Tripathi (1987), Khan and Wassilew (1987) and Khan *et al.* (1988) worked on the toxicity of neem oil and neem extracts against dermatophytes.

Tinea pedis is a chronic fungal infection, occurring in between the toes and characterized by shedding macerated skin, and is caused by *Trichophyton mentagrophytes*. Rai and Upadhyay *et al.* (1988) investigated the effect of stem bark on this disease and observed a significant growth inhibition.

The effect of petroleum ether extract of neem leaves on fungi was reported by Khan *et al.* (1991) while Iyer and Williamson (1991) reported that neem inhibited the protease activity of *Trichophyton* spp. Charles and Charles (1992) tried a mixture of fresh neem leaves and turmeric powder in the proportion of 4:1 by weight for the treatment of scabies caused by *Sarcoptes scabei*. It was a very safe, economical treatment and 94 percent of the patients were cured by it. A preparation with neem bark as the main ingredient was tried for leprosy by Subramanian and Lakshmanan (1993). The activity was comparable with modern anti-leprotic drugs. Srimathi and Murthy *et al.* (1993) studied a herbal preparation for scabies.

Conrick (1994) gave a brief account of skin diseases where neem can be used. Nat *et al.* (1991) in their review article after studying the literature, concluded that the immunostimulating property of neem may possibly be the reason for the recovery of patients with skin diseases.

USE IN VAGINITIS

Inflammation of the vagina due to genital tract infections in females is fairly common both in the developed and underdeveloped countries. The symptoms may vary from mild irritation to profuse leukorrhea and treatment varies from douching with whey (yogurt) to a broad range of antibiotic and antifungal treatment.

Garg *et al.* (1995) studied a purified fraction of neem in a cream base to assess the clinical efficacy of leukorrhea caused by *Chlamydia trachomatis* by applying cream vaginally one hour before going to bed. The formulation proved highly effective in clearing infection from the cervico-vaginal region. Upadhyaya and Dhawan (1994), when they treated supernatants of mouse spleen cells with neem, observed an inhibition of the intracellular multiplication of *Chlamydia*. Another polyherbal preparation widely tested for its antifertility effect as a cream and as a pessary by Talwar *et al.* (1995) had an antimicrobial effect against a wide range of pathogens of the vaginal canal, including *Candida albicans* and *Gardnerella vaginalis*. The cream was effective in the treatment of human papilloma 16 virus infecting the female genital tract. The other creams based on neem used by Usha Rani *et al.* (1994, 1995) were also effective for vaginal infections.

WOUND-HEALING (MUSCLE REGENERATION) PROPERTY

A polyherbal cream containing neem bark was tried by Bhatt and Koro (1984) on burning wound sepsis, bacteria and fungi. Thaker and Anjaria (1986) and Bhargava et al. (1986a) also found a wound-healing effect of neem, which according to Tandon et al. (1988) may be due to increased cutaneous capillary permeability at the site of the wound. Goget (1991) treated traumatic wound patients with neem fumigation.

Sathyanaryana et al. (1994) evaluated the wound-healing activity of neem oil ointment after 24 hours of wounding. At the end of the fourteenth day, the animals treated with 1 and 2 percent neem oil ointment showed complete epithelization. Hair growth was faster as compared to the other treatments and control. Sohni et al. (1995) studied septicemia in rats by using a traditional preparation containing neem, which was prophylactic against *Salmonella typhi*. When studied for safety (Tandon et al., 1995), neem oil was without any side effects on the liver and kidney. It was non-irritating to skin and safe for external use. It showed very low toxicity when given orally, with no histopathological changes.

REFERENCES

Abatan, M.O. and Makinde, M.J. (1986) Screening Azadirachta indica and Pisum sativum for possible antimalarial activities. *Journal of Ethnopharmacology*, **17**, 85–93.

Ahmad, I., Ahmad, F. and Hussain, S. (1995) In vitro anti-microbial activity of leaf and bark extract of Azadirachta indica A. Juss. *Indian Veterinary Medical Journal*, **19**, 204–206.

Anonymous (1979) Neem oil holds for diabetics. *Eastern Pharmacist*, **22**, 77.

Anonymous (1980) Neem useful in mouth cancer. *Drug and Pharmaceutical Industry Highlights*, **3**, 63.

Azambuja, P., Garcia, E.S., Ratcliffe, N.A. and Warthen, J.D. Jr. (1991) Immune depression in Rhodnius prolixus induced by the growth inhibitor, Azadirachtin. *Journal of Insect Physiology*, **37**, 771–777.

Babbar, O.P., Joshi, M.M. and Madan, A.R. (1982) Evaluation of plants for antiviral activity. *Indian Journal of Medical Research*, **12**, 54–65.

Badam, L., Deolankar, R.P., Kulkarni, M.M., Nagsampgi, B.A. and Wagh, W. (1987) In vitro antimalarial activity of neem (Azadirachta indica A. Juss) leaf and seed extract. *Indian Journal of Malariology*, **24**, 111–117.

Banerji, R. and Nigam, S.K. (1983) Anti-proteolytic activity of some triterpenoids. *International Journal of Crude Drug Research*, **21**, 93–95.

Barde, A.K. and Singh, S.M. (1983) Activity of plant extracts against Scytalidium anamorph of Hendersonula toruloidea, causing skin and nail diseases in man. *Indian Drugs*, **20**, 362–364.

Bardhan, J. (1991) Neem oil—a fertility controlling agent in rhesus monkey. *Indian Journal of Physiology and Pharmacology*, **35**, 278–280.

Benoit, F., Valentin, A., Pelissier, Y., Diafouka, F., Marion, C., Kone-Bamba, D., Mallie, M., Yapo, A. and Bastide, J.M. (1996) In vitro antimalarial activity of vegetal extracts used in West African traditional Medicine. *American Journal of Tropical Medicine and Hygiene*, **54**, 67–71.

Bhargava, A.K. (1986) A note on the use of neem (Azadirachta indica) oil as anti-hyperglycaemic agent in human volunteers of secondary diabetes. *Journal of Veterinary Physiology and Allied Sciences*, **5**, 45–48.

Bhargava, A.K. (1987) Neem oil as a synergist to anti-diabetic drugs for management of secondary hyperglycemia. *Neem Newsletter*, **4**, 31–32.

Bhargava, A.K., Dwivedi, S.K. and Singh, G.R. (1986) A note on the use of neem (Azadirachta indica) oil as anti-hyperglycaemic agent in dogs. *Indian Journal of Veterinary Science*, **6**, 66–67.

Bhargava, A.K., Lal, J., Sharma, A.K. and Kumar, P.N. (1986a) Neem oil (Azadirachta indica) an agent for promoting, healing of skin wounds. *Souvenir Convention ISVS*, 5–7 Nov. 1986, Haryana Agricultural University, Hissar, India.

Bhargava, K.P., Gupta, M.B., Gupta, G.P. and Mitra, C.R. (1970) Anti-inflammatory activity of saponins and other natural products. *Indian Journal of Medical Research*, **58**, 724–730.

Bhat, S.G. (1957) Antibacterial activity of some Indian inedible vegetable oils. *Indian Oilseed Journal*, **1**, 298–302.

Bhatt, R.M. and Koro, S. (1984) Clinical and experimental study of Panchvalkal and Shatavari on burn wound sepsis—bacterial and fungal. *Journal National Integrated Medical Association*, **26**, 131–133.

Bhide, N.K., Mehta, D.J. and Lewis, R.A. (1958) Diuretic action of sodium nimbidinate. *Indian Journal of Medical Science*, **12**, 141–145.

Bhide, N.K., Mehta, D.J., Attakar, M.W. and Lewis, R.A. (1958) Toxicity of sodium nimbidinate. *Indian Journal of Medical Science*, **12**, 146–148.

Bray, D.H., Conolly, J.D., Peters, W., Phillipson, J.D., Robinson, B.L., Tella, A., Thebtaranonth, Y., Warhurst, D.C. and Yuthavong, Y. (1985) Antimalarial activity of some limonoids. *Transaction of Royal Society of Tropical Medicine and Hygiene*, **79**, 426.

Bray, D.H., Warhurst, D.C., Connolly, J.D., O'Neill, M.J. and Phillipson, J.D. (1990) Plants as source of anti-malarial drugs. Part 7. Activity of some species of meliaceae plants and their constituent limonoids. *Phytotherapeutic Research*, **4**, 29–35.

Chakraborty, T. and Poddar, G. (1984) Herbal drugs in diabetes. Part I, Hypoglycaemic activity of indigenous plants in streptozotocin (STZ) induced diabetic rats. *Journal Institute of Chemists, India*, **56**, 20–22.

Chakraborty, T., Varotta, L. and Poddar, G. (1989) Evaluation of Azadirachta indica, leaf extract for hypoglycaemic activity in rats. *Phytotherapeutic Research*, **3**, 30–32.

Chattopadhyay, R.N., Maitra, S.K. and Chattopadhyay, R.R. (1993) Possible mechanism of anti-hyperglycaemic effect of *Azadirachta indica* leaf extract, Part I, *Fitoterapia*, **64**, 332–335.

Chattopadhyay, R.R., Chattopadhyay, R.N. and Maitra, S.K. (1993a) Possible mechanism of anti-hyperglycaemic effect of *Azadirachta indica* leaf extract. Part III, *Fitoterapia*, **64**, 535–538.

Chattopadhyay, R.R., Sarkar, S.K., Ganguly, S., Banerji, R.N., Basu, T.K. and Mukherjee, A. (1992) Hepatoprotective activity of Azadirachta indica leaves on paracetamol induced hepatic damage in rat. *Indian Journal of Experimental Biology*, **30**, 738–740.

Chattopadhyay, R.R., Sarkar, S.K., Ganguly, S. and Basu, T.K. (1994) A comparative evaluation of some anti-inflammatory agents of plant origin. *Fitoterapia*, **65**, 146–148.

Chattopadhyay, R.R., Sarkar, S.K., Ganguly, S. and Basu, T.K. (1993) Biochemical and toxicity studies with Azadirachta indica leaf extract. *Journal of Ecobiology*, **5**, 233–235.

Chatterjee, K.K. (1961) Treatment of cancer: a prelude. *Indian Medical Record*, **81**, 101.

Charles, V. and Charles, S.X. (1992) The use and efficacy of Azadirachta indica ADR (neem) and Curcuma longa (turmeric) in scabies. *Tropical and Geographical Medicine*, **44**, 178–181.

Chary, M.P., Reddy, E.J.S. and Reddy, S.M. (1984) Screening of indigenous plants for their anti-fungal principle. *Pesticides*, **18**, 17–18.

Chaudhry, R.R. and Haq, M. (1980) Review of plant screened for anti-fertility activity IV. *Bulletin of Medico-Ethno-Botanical Research*, **1**, 546–553.

Chopra, I.C., Gupta, K.C. and Nazir, B.N. (1952) Preliminary study of antibacterial substance from Melia azadirachta. *Indian Journal of Medical Research*, **40**, 511–515.

Cohen, E., Quistad, G.B. and Casida, J.E. (1996) Cytotoxicity of nimbolide, epoxyazadiradione, and other limonoids from neem insecticide. *Life Sciences*, **13**, 1075–1081.

Comley, J.C.W., Titanji, V.P.K., Ayafor, J.F. and Singh, V.K. (1990) In vitro antifilarial activity of some medicinal plants. *Acta Leidensia*, **59**, 361–363.

Conrick, J. (1994) *Skin care—the healing power of neem*. The Neem Association, Fl. 3351 (USA).

Dandiya, P.C. (1990) The pharmacological basis of herbal drugs acting on CNS. *Eastern Pharmacist*, **33**, 39–47.

David, S.N. (1965) The antifungal activity of neem oil and its constituents. *Mediscope*, **8**, 325.

David, S.N. (1969) Antipyretic activity of neem oil and its constituents. *Mediscope*, **12**, 25–27.

David, S.N. (1978) Effect of neem oil and its constituents on cotton pellet inflammation. *Mediscope*, **20**, 273–274.

Debelmas, A.M. and Hache, J. (1976) Etude pharmacologique de quelques plantes medicinales du Nepal. Toxicite 'aigue, etude compartementale et action sur le system nerveux central. *Plantes Medicinales et Phytotherapie*, **10**, 18–138.

Deshpande, V.Y., Mendulkar, K.N. and Sadre, N.L. (1980) Male antifertility activity of Azadirachta indica in mice—a preliminary report. *Journal of Postgraduate Medicine*, **26**, 167–170.

Devdas, K.V., Radhakrishnan, P. and Warrier, P.K. (1995) Effect of nimbidin in the management in parinamsula. *Seminar Central Council for Research in Ayurveda and Siddha*, 20 March, 1995, New Delhi, India, pp. 41.

Dixit, V.P., Jain, P., Purohit, A.K., Jain, P. and Raychaudhuri, S.P. (1992) Medicinal uses of neem (Azadirachta indica) in fertility regulation, diabetes, and atherosclerosis. *Recent Advances in Medicinal, Aromatic and Spice Crops, Volume 2*. International Conference held on 28–31 January 1989 at New Delhi, India, pp. 463–471.

Dixit, V.P., Sinha, R. and Tank, R. (1986) Effect of neem seed oil in the blood glucose concentration of normal and alloxan diabetic rats. *Journal of Ethanopharmacology*, **17**, 95.

Dube, S. and Tripathi, S.C. (1987) Toxicity of some plants against dermatophytes. *National Academy of Sciences, India, Science Letter*, **10**, 45–48.

Ekanem, J. (1976) Dongo-Yaro. Does it work? *Nigerian Medical Journal*, **8**, 8–10.

El-Hawary, Z.M. and Khalief, T.S. (1990) Biochemical studies on hypoglycaemic agents (1) Effect of Azadirachta indica. *Archives Pharmaceutical Research*, **13**, 108–112.

Fabry, W., Okemo, P. and Ansorg, R. (1996) Fungistatic and fungicidal activity of east African medicinal plants. *Mycoses*, **39**, 67–70.

Fujiwara, T., Takeda, T., Ogihara, M., Shimizu, M., Nomura, T. and Tomita, Y. (1984) Further studies on the structure of polysaccharides from the bark of Melia azadirachta. *Shoyakagaku Zasshi*, **38**, 3343–3403.

Fujiwara, T., Takeda, T., Ogihara, M., Shimizu, M. and Tomita, Y. (1982) Studies on the structure of polysaccharides from the bark of Melia azadirachta. *Chemical and Pharmaceutical Bulletin, Tokyo*, **30**, 4025–4030.

Gandhi, M., Lal, R., Sankaranaryanan, A., Banerjee, C.K. and Sharma, P.L. (1988) Acute toxicity study of the oil from Azadirachta indica seed (Neem oil). *Journal of Ethnopharmacology*, **23**, 39–51.

Garcia, E.S., Gonzalez, M.S., Azambuja, P. and Rembold, H. (1989) Chagas' disease and its insect vector. Effect of azadirachtin A on the interaction of a triatomine host (Rhodnius prolixus) and its parasite (Trypanosoma cruzi). *Zeitschrift fur Naturforschung. C. Biosciences*, **44**, 317–322.

Garg, A.K., Agarwal, D.K., Singh, S.D. and Nath, K. (1992) Growth and spermatogenesis in rats, fed solvent extracted water washed and untreated neem (Azadirachta indica A. Juss) seed kernel cake. *International Journal of Animal Science*, **7**, 223–225.

Garg, G.P., Nigam, S.K. and Ogle, C.W. (1993) The gastric anti ulcer effect of the leaves of neem tree. *Planta Medica*, **59**, 215–217.

Garg, S., Singh, R., Kaur, R., Dhar, V., Taluja, V. and Talwar, G.P. (1995) (Personal communication) Praneem polyherbal cream for contraception and genital tract infection. *National Institute of Immunology, New Delhi*.

Garg, S., Talwar, G.P. and Upadhyay, S.N. (1994) Comparison of extraction procedures on the immunocontraceptive activity of neem seed extracts. *Journal of Ethnopharmacology*, **44**, 87–92.

Garg, S., Talwar, G.P., Upadhyay, S.N., Mittal, A. and Kapoor, S. (1993a) Identification and characteristic of the immuno-modulater fraction from neem seed extract responsible for long term antifertility activity after intra-uterine administration (abstr.). *Proceed. World Neem Conference*, 24–28 Feb., 1993, Bangalore, India.

Gogate, S.S. and Marathe, A.D. (1989) Antiviral effect of neem leaf (Azadirachta indica A. Juss) extract on Chikungunya and Measles viruses. *Journal of Research and Education in Indian Medicine*, **8**, 1–5.

Goget, R.B. (1991) Traumatic wound patients and Dhoopan chikitsa. *Deerghayu International*, **7**, 2–8.

Gopal Raj, N. (1993) Neem boosts body's defense. *Neem News*, **1**, 1.

Grand, A.-Le. (1989) Anti-infectious phytotherapies of the tree savanna. Senegal (West Africa) III: a summary of the phytochemical substances and the anti-microbal activity of 43 species. *Journal of Ethnopharmacology*, **25**, 315–338.

Handa, S.S., Chawala, A.S. and Maninder (1989) Hypoglycaemic plants—a review. *Fitoterapia*, **60**, 200.

Handa, S.S., Chawala, A.S. and Sharma, A.K. (1992) Plants with anti-inflammatory activity. *Fitoterapia*, **63**, 3–31.

Handa, S.S., Sharma, A. and Chakraborty, K.K. (1986) Natural products and plants as liver protecting drugs. *Fitoterapia*, **57**, 307–351.

Ilesanmi, O.R, Aladesanmi, A.J. and Adeoye, A.O. (1988) Pharmacological investigation on the cardiac activity of some Nigerian medicinal plants. *Fitoterapia*, **59**, 371–376.

Iwalewa, E.O., Lege Oguntoye, L., Rai, P.P., Iyaniwura, T.T. and Etkin, N.L. (1995) Effect of cardiogenic glycoside and sterol extracts of Azadirachta indica leaves on Plasmodiuim yoelii nigeriensis infection in mice. *International Conference on Current Progress in Medicinal and Aromatic Plants Research*, 30 Dec. 1994, Calcutta (India).

Iwu, M.M., Obidoa, O. and Anazodo, M. (1986) Biochemical mechanisms of the anti-malarial activity of Azadirachta indica leaf extract. *Pharmacological Research Communication*, **18**, 81–91.

Iyer, S.R. and Williamson, D. (1991) Efficacy of some plant extract to inhibit the protease activity of Trichophyton species. *Geobios*, Jodhpur, **18**, 3–6.

Jacobson, M. (1989) Pharmacology and toxicology of neem. In Jacobson, M. (ed.) *Focus on Phytochemical Pesticides*. Vol. I. *The Neem Tree*. CRC Press Inc., Florida, USA.

Jain, J.P. and Pathak, V.N. (1970) Antifungal activity in leaf extracts of certain plants. *Labdev Journal of Science and Technology*, **88**, 58.

Jain, P.P., Suri, R.K., Deshmukh, S.K. and Mathur, K.C. (1987) Fatty oils from oilseeds of forest origin as antibacterial agent. *Indian Forester*, **113**, 297–299.

Jaiswal, A.K., Bhattacharya, S.K. and Acharya, S.B. (1994) Anxiolytic activity of Azadirachta indica leaf extract in rats. *Indian Journal of Experimental Biology*, **32**, 489–491.

Jones, I.W., Denholm, A.A., Lev, S.V., Lovell, H., Wood, A. and Sinden, R.E. (1994) Sexual development of malaria parasite is inhibited in vitro by the neem extract azadirachtin and its semi-synthetic analogues. *FEMS Microbiology Letters*, **120**, 267–274.

Joshi, C.G. and Nagar, N.G. (1952) Antibiotic activity of some Indian medicinal plants. *Journal of Scientific and Industrial Research*, **11B**, 261.

Kasturi, M., Manivannan, B., Ahamed, R.N., Shaikh, P.D. and Pathan, K.M. (1995) Changes in epididymal structure and function of albino rat treated with Azadirachta indica leaves. *Indian Journal of Experimental Biology*, **33**, 725–729.

Khalid, S.A. and Duddeck, H. (1989) Isolation and characterisation of an antimalarial agent of the neem tree Azadirachta indica. *Journal of Natural Products*, **52**, 922–926.

Khalid, S.A., Farouk, A., Geary, T.G. and Jensen, J.B. (1986) Potential anti-malarial candidate from African plants. An in vitro approach using Plasmodium falciparum. *Journal of Ethnopharmacology*, **14**, 45–51.

Khan, M., Schneider, B., Wassilew, S.W. and Splanemann, V. (1988) The effect of raw materials of the neem tree, neem oils and neem extracts on dermatophytes, yeasts and moulds (in German). *Zeitschrift fur Hautkrankheiten*, **63**, 499–502.

Khan, R.M. and Wassilew S.W. (1987) The effect of raw material from the neem tree, neem oil and neem extract on fungi pathogenic to humans. Natural pesticides from the neem tree (Azadirachta indica A. Juss) and other tropical plants. *Proc. of the 3rd Int. Neem Conference*, Nairobi, Kenya, 10–15 July, 1986.

Khan, R.M., Plempei, M., Wassilew, S.W. and Raychaudhuri, S.P. (1991) The effect of the petroleum ether extract of neem leaves on fungi pathogenic to humans in vitro and in vivo. *Recent Advances in Medicinal, Aromatic and Spice Crops* (Vol. I). International Conference held on 28–31 January, 1989 at New Delhi, India, 269–272.

Khanna, N., Goswami, N., Sen, P. and Ray, A. (1995) Antinociceptive action of Azadirachta indica (neem) in mice. Possible mechanism involved. *Indian Journal of Experimental Biology*, **33**, 848–850.

Khatak, S.G., Gilani, S.N. and Ikram, M. (1985) Antipyretic studies on some indigenous Pakistani medicinal plants. *Journal of Ethnopharmacology*, **14**, 45–51.

Kher, H.N., Dave, M.R. and Dholakia, P.M. (1984) In vitro anti-bacterial activity of chloroform extract of Azadirachta indica leaves. *Gujvet*, **13**, 17–19.

Koley, K.M., Lal, J. and Tandon, S.K. (1994) Anti-inflammatory activity of Azadirachta indica (neem) leaves. *Fitoterapia*, **65**, 524–528.

Komolafe, O.O., Anyabluke, C.P., Obaseki, A.O. (1988) The possible role of mixed function oxidases in the hepatobiliary toxicity of Azadirachta indica. *Fitoterapia*, **59**, 109–113.

Kroes, B.H., Berg, A.J.J. van den, Heystra, E.A., De Silva, K.T.D. and Labadie, R.P. (1990) Fermentation in traditional medicine: the impact of Woodfordia fruticosa flowers on the immunomodulater activity and the alcohol and sugar contents of Nimba Arishta (a potion of fermented Azadirachta indica bark). *Planta Medica*, **56**, 667.

Kroes, B.H., Berg, A.J.J. van den, Labadie, R.P., Abeysekera, A.M. and De Silva, K.T.D. (1993) Impact of the preparation process on immunomodulatory activities of the ayurvedic drug Nimba arishta. *Phytotherapy Research*, **7**, 35–40.

Kroes, B.H., Berg, A.J.J. van den, Queries, van Ufford, H.C., Dijk, H. van and Labadie, R.P. (1992) Anti-inflammatory activity of gallic acid. *Planta Medica*, **58**, 499–504..

Labadie, R.P., Nat, J.M. van der, Simons, J.M., Kroes, H., Kosasi, S., Berg, A.J.J. van den, tHart, L.A., Sluis, W.G., van der, Abeysekera, A., Bamunuarachchi and De Silva, K.T.D. (1989). Ethnopharmacology approach to the search for immunomodulaters of plant origin. *Planta Medica*, **55**, 339–348.

Lal, R., Sankarnaryanan, A., Mathur, V.S. and Sharma, P.L. (1986) Anti-fertility effect of neem oil in female albino rats by the intra-vaginal and oral routes. *Indian Journal of Medical Research*, **83**, 89–92.

Lal, R., Gandhi, M., Sankarnaryanan, A., Mathur, V.S. and Sharma, P.L. (1987) Antifertility effect of Azadirachta indica oil administered per os to female albino rats on selected days of pregnancy. *Fitoterapia*, **58**, 230–242.

Latiff, A. (1979) Formulation for treating hair. British UK Pat. Appl. 2000971, 24 Jan., 1979, vide *Chemical Abstract*, 92, 11100H, 1980.

Lorenz, H.K.P. (1976) Neem tree bark extract in the treatment of inflammatory stomatis. *Zahnaerztle Prasas*, **8**, 1–4.

Luscombe, D.K. and Taha, S.A. (1974) Pharmacological studies on the leaves of Azadirachta indica. *Journal of Pharmacy and Pharmacology*, **26**, Suppl., 111P.

Mahajan, V.M. (1986) Further studies on antimycotic agents. *Mykosen*, **29**, 407–412.

Masilamani, G., Bhardwaj, T.R.P. and Purushothaman (1979) Role of Eth Ennei and Nanguneri Enni in the treatment of 'Karappan' (Eczema)—pilot study. *Journal for Research in Indian Medicine, Yoga and Homeopathy*, **14**, 74–80.

Mateenuddin, M., Kharaitkar, K.K., Mendhulkar, K.N. and Sadre, N.L. (1986) Assessment of estrogenicity of neem (Azadirachta indica) leaf extract in rats. *Indian Journal of Physiology and Pharmacology*, **30**, 118–120.

Mukherji, B. (1957) Indigenous Indian drugs used in the treatment of diabetes. *Journal of Scientific and Industrial Research*, **16A**, 1–18.

Mukherji, S., Talwar, G.P. and Lohiya, N.K. (1995) Termination of pregnancy in rodents and primates by oral administration of Praneem—a purified neem seed extract. *International Symposium on Prospects of Zona pellucida Glycoprotein for Immunoconception and 7th Annual Conference of Indian Society for Study of Reproduction and Fertilization*. National Institute of Immunology, New Delhi, p. 128.

Mukherji, S., Garg, S. and Talwar, G.P. (1995a) Termination of pregnancy by Praneem' a purified fraction of neem (Azadirachta indica). *International Conference on Current Progress in Medicinal and Aromatic Plant Research*, Calcutta, India, pp. 142–143.

Murty, K.S., Rao, D.N., Rao, D.K. and Murty, L.P.G. (1978) A preliminary study on hypoglycaemic and anti-hyperglycaemic effect of Azadirachta indica. *Indian Journal of Pharmacology*, **10**, 247–250.

Murthy, S.P. and Sirsi, M. (1957) Pharmacological studies on Melia Azadirachta. Part 1. Antibacterial, antifungal, and antitubercular activity of neem oil and its fractions. *Symposium on Utilization of Medicinal Plants*, Lucknow, India.

Murthy, S.P. and Sirsi, M. (1958) Pharmacological studies on Melia azadirachta L. *Indian Journal of Physiology and Pharmacology*, **2**, 387–396.

Murthy, S.P. and Sirsi, M. (1958a) Pharmacological studies in Melia azadirachta. Part II. Estrogenic and antipyretic activity of neem oil and its fractions. *Indian Journal of Physiology and Pharmacology*, **2**, 456.

Murthy, S.P. and Sirsi, M. (1958b) Pharmacological studies on neem oil on experimental Avian malaria (P. gallinaceum). *Journal Mysore Medical Association*, **23**, 1.

Nadkarni, K.M. (1954) *Indian Materia Medica*. Revised by A.K. Nadkarni, reprint (1976). Popular Prakashan, Bombay, India.

Nair, P.R., Namboodiri, M.N.S., Madhavikutty, P. and Prabhakaran, V.A. (1987) Clinical evaluation of Ayurvedic preparation in vitiligo. *Journal for Research in Ayurveda and Siddha*, **8**, 20–38.

Narayana, D.S. (1965) The anti fungal activity of neem oil and its constituents. *Mediscope*, **8**, 326.

Nat, J.M. van der, tHart, L.A., Sluis, W.G. van der, Dijk, H. van, De Silva, K.T.D. and Labadie, R.P. (1989) Characterisation of anti-complement compounds from Azadirachta indica. *Journal of Ethnopharmacology*, **27**, 15–14.

Nat, J.M. van der, Klerx, J.P.A.M., Dijk, H. van, De Silva, K.T.D. and Labadie, R.P. (1987) Immunomodulatory activity of aqueous extract of Azadirachta indica stem bark. *Journal of Ethnopharmacology*, **19**, 125–131.

Nat, J.M. van der, Sluis, W.G. van der, De Silva, K.T.D. and Labadie, R.P. (1991) Ethnopharmacognostical survey of Azadirachta indica A. Juss (Meliaceae). *Journal of Ethnopharmacology*, **35**, 1–24.

Nat, J.M. van der, Sluis, W.G. van der, tHart, L.A., Dijk, H. van, De Silva, K.T.D. and Labadie, R.P. (1991a) Activity guided isolation and identification of Azadirachta indica bark extract constituents which specifically inhibit chemiluminescence production by activated human polymorphonuclear leukocytes. *Planta Medica*, **57**, 65–68.

Nok, A.J., Esieva, K.A.N., Longdet, I., Arowosafe, S., Onyenekwe, P.C., Gimba, C.E. and Kagtu, J.A. (1993) Trypanocidal potentials of Azadirachta indica: in vivo activity of leaf extract against Trypansoma brucei brucei. *Journal of Clinical Biochemistry and Nutrition*, **15**, 113–118.

Obaseki, A.O., Adeyi, O. and Anyabuika, C. (1985) Some serum enzyme level as marker of possible acute effects of the aqueous extract of Azadirachta indica on membranes in vivo *Fitoterapia*, **52**, 111–115.

Obaseki, A.O. and Jegde-Fadunsin, H.A. (1986) Anti-malarial activity of Azadirachta indica. *Fitoterapia*, **57**, 247–251.

Obih, P.O. and Makinde, J.M. (1985) Effect of Azadirachta indica on Plasmodium berghei in mice. *African Journal of Medical Sciences*, **14**, 51–54.

Okpako, D.T. (1977) Prostaglandin synthetase inhibitory effect of Azadirachta indica. *Journal of West African Science Association*, **22**, 45–47.

Okpanyi, S.N. and Ezeukwu, G.C. (1981) Anti-inflammatory and anti-pyretic activity of Azadirachta indica. *Planta Medica*, **41**, 34–39.

Oomacham, M., Srivastava, J.L. and Masih, S.K. (1990) Observation on certain plants used in the treatment of jaundice. *Indian Journal of Applied and Pure Biology*, **5**, 99–102.

Paranjape, M.H. and Paranjape, M.M. (1993) Use of neem oil Azadirachta indica suppositories as contraceptive. *Proc. World Neem Conference*, 26–28 Feb., 1993, Bangalore, India.

Parshad, O., Singh, P., Gardner, M., Fletcher, C., Rickards, E. and Choo Kang, E. (1994) Effect of aqueous neem extract (Azadirachta indica) on testosterone and other blood constituents in male rats. *West Indian Medical Journal*, **43**, 71–74.

Patel, R.P. and Trivedi, B.M. (1962) The in vitro antibacterial activity of some medicinal oils. *Indian Journal of Medical Research*, **50**, 212–222.

Pettit, G.R., Barton, D.H.R., Herald, C.L., Polonsky, J., Schimdt, J.M. and Connolly, J.D. (1983) Evaluation of limonoids against the murine P388 lymphocytic leukemia cell line. *Journal Natural Products*, **46**, 379–380.

Phillipson, J.D. and O'Neill, M.J. (1989) New leads to the treatment of protozoal infections based on natural product molecules. *Acta Pharmaceutica Nordica*, **1**, 131–141.

Pillai, N.R. (1978) Anti-gastric ulcer activity of nimbidin. *Indian Journal of Medical Research*, **68**, 169–175.

Pillai, N.R. (1995) Nimbtiktam—a potential anti-ulcer drug. *Seminar on Research in Ayurveda and Siddha*, New Delhi, India, 20–22 March, 1996, pp. 63.

Pillai, N.R. and Santhakumari, G. (1981) Anti-arthritic and anti-inflammatory actions of nimbidin. *Planta Medica*, **43**, 59–63.

Pillai, N.R. and Santhakumari, G. (1981a) Hypoglycaemic activity of Melia azadirachta Linn (Neemb) (abstr.). *Indian Journal of Pharmacology*, **13**, 91–92.

Pillai, N.R. and Santhakumari, G. (1981b) Hypoglycaemic activity of Melia azadirachta Linn (Neem). *Indian Journal of Medical Research*, **74**, 931–933.

Pillai, N.R. and Santhakumari, G. (1984) Effect of nimbidin on acute and chronic gastro-duodenal ulcer models in experimental animals. *Planta Medica*, **50**, 143–146.

Pillai, N.R. and Santhakumari, G. (1984a) Toxicity studies on nimbidin, a potential anti ulcer drug. *Planta Medica*, **50**, 146–148.

Pillai, N.R., Suganthan, D. and Santhakumari, G. (1980) Analgesic and antipyretic action of nimbidin. *Bulletin Medico Ethnobotanical Research*, **1**, 393–400.

Prakash, A.O. (1986) Potentialities of some indigenous plants for anti-fertility activity. *International Journal of Crude Drug Research*, **24**, 19–24.

Prakash, A.O., Mishra, A. and Mathur, R. (1991) Studies on the reproductive toxicity due to extract of Azadirachta indica (seeds) in adult cycle female rats. *Indian Drugs*, **28**, 163–169.

Prakash, A.O., Mishra, A., Metha, H. and Mathur, R. (1991a) Effect of ethanolic extract of Azadirachta indica seeds on organs in female rats. *Fitoterapia*, **62**, 99–105.

Prakash, A.O., Shukla, S. and Mathur, R. (1987) Interceptive plants: present status and future aspects. *Comparative Physiology and Ecology*, **12**, 157–171.

Prakash, A.O., Shukla, S. and Mathur, R. (1988) Non-hormonal post coital contraceptive action of neem oil in rats. *Journal of Ethnopharmacology*, **23**, 53–59.

Prasad, H.C., Mazumdar, R. and Chakraborty, R. (1993) Research on two medicinal plants from Ayurvedic system of medicine Azadirachta indica A. Juss (syn. Melia azadirachta) and Melia azedarach Linn—their, past, present and future. *Deerghayu International*, July–Sept., 24–28.

Qadri, S.S.H., Usha, G. and Jabaan, K. (1984) Sub acute dermal toxicity of Neemrich-100 (tech.) to rats. *International Pest Control*, 26, 18–20.

Rai, A. and Sethi, M.S. (1972) Screening of some plants for their activity against vaccinia and fowl pox viruses. *Indian Journal of Animal Science*, 42, 1066–1070.

Rai, M.K. (1988) In vitro sensitivity of Microsporum nanum to some plant extracts. *Indian Drugs*, 25, 521–523.

Rai, M.K. and Upadhyay, S. (1988) Screening of medicinal plants of Chhindwara District against Trichophyton mentagrophytes: a casual organism of tinea pedis. *Hindustan Antibiotics Bulletin*, 30, 33–36.

Rajasekharan, S., Pillai, N.G.K.P., Kurup, P.B., Pillai, K.G.B. and Nair, C.P.R. (1980) Effect of nimbidin in psoriasis—a case report. *Journal of Research in Ayurveda and Siddha*, 1, 52–58.

Ramaniyalu, G. (1986) In vitro antibacterial activity of neem oil. *Indian Journal of Medical Research*, 34, 314–316.

Ramaswamy, A.S. and Sirsi, M. (1967) Anti-tubercular activity of some natural products. *Indian Journal of Pharmacy*, 29, 157–159.

Rao, A.R., Kumar, S.S., Paramasivam, T.B., Kamalkashi, S., Parashuram, A.R. and Shantha, M. (1969) Study of antiviral activities of tender leaves of margosa tree (Melia azadirachta) on vaccinia and variola viruses—a preliminary report. *Indian Journal of Medical Research*, 57, 495–502.

Rao, D.C.P. and Satyanaryana, T. (1977) Antibacterial activity of some medicinal plant extracts. *Indian Drugs and Pharmaceutical Industry*, 12, 21–22.

Rao, D.V.K., Chopra, P., Chabra, P.C. and Ramanvualu, G. (1986) In vitro anti bacterial activity of neem oil. *Indian Journal of Medical Research*, 34, 314–316.

Reddy, A.B. and Sethi, M.S. (1984) Preliminary studies on the antiviral effects of some indigenous plant extracts (Berberis floribunda, Azadirachta indica, Moringa oleifera, Cynodon dactylon) on vaccinia virus in vivo. *Annali della Facalta di Medicina Veterinaria Messina*, 21, 65–76.

Reddy, M.K., Kokate, C.K. and Chari, N. (1984a) Anti ovulatory effect of different crude drug combinations in female albino rats. *Ancient Science of Life*, 4, 132–134.

Reddy, M.K., Ravi, A., Kokate, C.K. and Chari, N. (1984b) Effect of some crude drug combinations on oestrus cycle in albino rats. *Eastern Pharmacist*, 27, 139–140.

Riar, S.S., Bardhan, J., Thomas, P., Kain, A.R. and Rajinder Parshad (1988) Mechanism of antifertility action of neem oil. *Indian Journal of Medical Research*, 88, 339–342.

Rochanakij, S., Thebtaranonth, Y., Yenjai, C.H. and Yuthavong, Y. (1985) Nimbolide, a constituent of Azadirachta indica inhibits Plasmodium in culture. *Southeast Asian Journal of Tropical Medicine and Public Health*, 16, 66–72.

Salunkhe, D.K. and Adsale, R.N. (1985) Use of plants to control human fertility. *Current Research Report*, 1, 113–117.

Sampathraj, P., Badri S. and Vanithakumari, G. (1993) Effect of neem oil: structural and functional changes in the epididymis of rats. *Proc. World Neem Conference*, 24–28 Feb., 1993, Bangalore, India.

Sathyanaryana, D., Mohan Kumar, V., Sarvanan, K. and Dandabany, D. (1994) Evaluation of wound healing activity of neem oil ointment (abstr.). *46th Indian Pharm. Congress*, 28–30 Dec., 1994, Chandigarh (India), pp. 384.

Satyanaryana, T. and Rao, D.P.C. (1977) Activity of some medicinal plant extracts against keratinophilic fungi. *Indian Drugs & Pharmaceutical Industry*, 11, 7–8.

Sen, P., Mediratta, A. and Puri, S. (1993) An experimental evaluation of Azadirachta indica (neem) in normal and stressed rats and adaptogenic effects. *Proc. World Neem Conference*, 24–28 Feb., 1993, Bangalore, India.

Shah, M.P., Sheth, U.K., Bhide, N.K. and Shah, M.J. (1958) Clinical trials with parenteral sodium nimbidinate in a new diuretic. *Indian Journal of Medical Sciences*, 12: 150.

Shaikh, P.D., Manivannan, B., Pathan, K.M., Kasturi, M., Ahmed, R.N. and Nazeer Ahamed, R. (1993) Antispermatic activity of Azadirachta indica leaves in albino rats. *Current Science*, 64, 688–689.

Shankaranaryana, D. (1978) Effect of neem oil and its constituents on cotton pellet inflammations. *Mediscope*, **20**, 273–274.

Sharma, J.D., Jha, R.K., Gupta, I., Jain, P. and Dixit, V.P. (1987) Anti-androgenic properties of neem seed oil (Azadirachta indica) in male rats and rabbits. *Ancient Science of Life*, **7**, 30–38.

Sharma, M.K., Khara, A.K. and Hasan Feroze (1983) Effect of neem oil in blood sugar levels of normal, hyperglycaemic and diabetic animals. *Nagarjun*, **26**, 247–250.

Sharma, V.N. and Saksena, K.P. (1959) Spermicidal action of sodium nimbinate. *Indian Journal of Medical Research*, **47**, 322–334.

Sharma, V.N. and Saksena, K.P. (1959a) Sodium nimbinate: *In vitro* study of its spermicidal action. *Indian Journal of Medical Science*, **13**, 10–38.

Shilaskar, D.V. and Parashar, G.C. (1989) Evaluation of indigenous anthelmintics. In vitro screening of some indigenous plants for their anthelmintic activity against Ascardia galli. *Indian Journal of Indigenous Medicine*, **6**, 49–53.

Shrivastava, A.K. and Singh, K.V. (1982) Attempts to understand the effect of essential oils obtained from onion, garlic and neem leaves against different dermatophytes. *Indian Drugs*, **19**, 245.

Shukla, R., Singh, S. and Bhandari, C.R. (1973) Preliminary clinical trials on antidiabetic action of Azadirachta indica. *Medicine and Surgery*, **13**, 11–12.

Siddiqui, S., Shaheen, F., Siddique, B.S. and Ghiasuddin (1992) Constituents of Azadirachta indica, isolation and structure elucidation of a new antibacterial tetranortriterpenoid, mahmoodin, and a new protolimonoid, naheedin. *Journal of Natural Products*, **55**, 303–310.

Singh, K. and Singh, D.K. (1996) Mollusicidal activity of neem (Azadirachta indica). *Journal of Ethnopharmacology*, **24**, 35–40.

Singh, L. and Sharma, M. (1978) Antifungal properties of some plant extract. *Geobios*, **5**, 49–53.

Singh, N., Misra, N., Singh, S.P. and Kohli, R.P. (1979) Melia azadirachta in some common skin disorders—a clinical evaluation. *The Antiseptic*, **76**, 677–679.

Singh, N., Nath, R., Singh, S.P. and Kohli, R.P. (1980) Clinical evaluation of anthelimintic activity of Melia Azadirachta. *The Antiseptic*, **77**, 739–741.

Singh, N. and Sastry, M.S. (1981) Antimicrobial activity of neem oil (abstr.). *Indian Journal of Pharmacology*, **13**, 102.

Singh, P.P., Junnarakar, A.Y., Reddi, G.S and Singh, K.V. (1987) Azadirachta indica: neuropsychopharmacological antimicrobial studies. *Fitoterapia*, **58**, 235–238.

Singhal, K.K. and Mudgal, V.D. (1983) Versatile neem (Azadirachta indica) a review. *Agricultural Review*, **4**, 1–10.

Sinha, K.C., Riar, S.S., Jaya, B., Thomas, P., Jain, A.K. and Jain, R.K. (1984) Anti-implantation effect of neem oil. *Indian Journal of Medical Research*, **80**, 708–710.

Sinha, K.C., Riar, S.S., Tiwary, R.S., Dhawan, A.K., Jaya, B., Thomas, P., Jain, A.K. and Jain, R.K. (1984a) Neem oil as a vaginal contraceptive. *Indian Journal of Medical Research*, **79**, 131–136.

Sohni, Y.R., Kaimal, P. and Bhatt, R.M. (1995) Prophylactic therapy of Salmonella typhi septicemia in mice with a traditionally prescribed crude drug formulation. *Journal of Ethnopharmacology*, **45**, 141–147.

Souza, C. de, Amegavi, K.K., Kourmaglo, K. and Gbeassor, M. (1993) Study of the antimicrobial activity of the total aqueous extracts of ten medicinal plants. *Revue de Medecines et Pharmacopees Africaines*, **7**, 109–115.

Srimathi, Y.B. and Murthy, Uma, D. (1993) Studies on the efficacy of herbal formulation in wound healing and scabies. *45th Indian Pharmaceutical Congress*, New Delhi, India, pp. 34.

Subramanian, M.S. and Lakshmanan, K.K. (1993) Azadirachta indica A. Juss. neem drugs, as an antileprosy source. *Proc. World Neem Conference*, 24–28 Feb. 1993, Banglore, India.

Takeya, K., Qiao, Z.S., Hirobe, C. and Itokawa, H. (1996) Cytotoxic azadirachtin type limonoids from Melia azadarach. *Phytochemistry*, **42**, 709–712.

Talwar, G.P., Garg, S., Dhar, V., Chabra, R., Ganju, A. and Upadhyaya, S.N. (1995) Praneem Polyherbal cream and pessaries with dual properties of contraception and alleviation of genital infection. *Current Science*, **68**, 437–440.

Tandon, S.K., Chandra, S., Gupta, S., Tripathi, H.C. and Lal, J. (1990) Pharmacological effects of Azadirachta indica leaves. *Fitoterapia*, **61**, 75–78.

Tandon, S.K., Gupta, S., Chandra, S. and Lal, J. (1988) Increasing action of vascular permeability by Azadirachta indica seed oil (neem oil). *Indian Journal of Pharmacology*, **20**, 203–205.

Tandon, S.K., Gupta, S., Chander, S., Lal, J. and Rajinder Singh (1995) Safety evaluation of Azadirachta indica seed oil a herbal wound dressing agent. *Fitoterapia*, **66**, 69–72.

Tella, A.(1976) Studies on Azadirachta indica in malaria. *British Journal of Pharmacology*, **58**, 318.

Terumo Corporation Patent (1983) Anti-inflammatory polysaccharide from Melia azadirachta. Japan Kokai Tokyo KOHO JP 8205532 (83 225021) pp, *vide Chemical Abstract*, **100**, 91350n.

Terumo Corporation Patent (1985) Preparation of anti-tumour polysaccharide N9Gl from Melia azadirachta. Japan Kokai Tokyo Koho JP 60 19718 (8519718) 10 pp. *vide Chemical Abstract*, **103**, 11440q.

Tewari, R.K., Mathur, R. and Prakash, A.O. (1986) Postcoital antifertility effect of neem oil in female albino rats. *International Research Communication System of Medical Sciences*, **14**, 1005–1006.

Thaker, A.M. and Anjaria, J.V. (1986) Antimicrobial and infected would healing response of some traditional drugs. *Indian Journal of Pharmacology*, **18**, 171–174.

Thakur, R.S., Puri, H.S. and Akhtar Hussain (1989) *Major Medicinal Plants of India*. Central Institute of Medicinal and Aromatic Plants, Lucknow, India, pp. 85–91.

Thind, T.S. and Dahiya, M.S. (1977) Inhibitory effects of essential oils of four medicinal plants against keratinophilic fungi. *Eastern Pharmacist*, **20**, 147–148.

Thompson, E.B. and Anderson, C.C. (1978) Cardiovascular effects of Azadirachta indica extract. *Journal of Pharmaceutical Sciences*, **67**, 1476–1478.

Tidjani, M.A., Dupont C. and Wepierre (1989) Azadirachta indica stem bark extract. Anti-inflammatory activity. *Plants Medicinal et Phytotherapy*, **23**, 259–266.

Tyagi, R.K. and Sahrawat, D. (1995) Studies on the contraceptive effect of neem oil. *Seminar on Research in Ayurveda and Siddha*, 20–22 March, 1995. Central Council for Research in Ayurveda and Siddha, Delhi (India), p. 5.

Tyagi, R.K., Tyagi, M.K., Goyal, H.R. and Sharma, K. (1977) Anthelimintic effect of an Ayurvedic compound in cases of ankushkrimi (ankylostomiasis). *Nagarjun*, **21**, 5–7.

Udeinya, I.J. (1993) Anti-malarial activity of Nigerian neem leaves. *Transaction of the Royal Society of Tropical Medicine and Hygiene*, **87**, 471.

Upadhyay, R.K. and Arora, D.K. (1975–76) Sporostatic nature of neem smoke and its possible ecological influence on air fungal flora of a polluted site. *Journal Scientific Research, Banaras Hindu University, India*, **26**, 125–129.

Upadhyay, V.P. (1984) Diabetes mellitus and its management by indigenous resources. *Nagarjun*, **28**, 6–8.

Upadhyaya, S. and Dhawan, S. (1994) Neem (Azadirachta indica) Immunomodulater properties and therapeutic potential. *Update Ayurveda*, 24–26 Feb., 1994, Bombay, India, pp. 34.

Upadhyaya, S.N., Dhawan, S., Garg, S., Wali, N., Tucker Lynn and Anderson, D.J. (1993) Immunomodulatery properties of neem (Azadirachta indica). (abstr.) *Proc. World Neem Conference*, 24–28 February, 1993, Bangalore, India.

Upadhyaya, S.N., Kaushic, C. and Talwar, G.P. (1990) Anti-fertility activity of Neem (Azadirachta indica) oil by single intra-uterine administration. A novel method for contraception. *Proceedings Royal Society, London*, **B 242**, 175–179.

Usha Rani, P., Naidu, M.U.R., Ramesh Kumar, T. and Shobha, U. (1994) Evaluation of new herbal ointment in vaginitis. *Update Ayurveda*, 24–26 February, 1994, Bombay, India, pp. 77.

Usha Rani, P., Naidu, M.U.R., Rama Raju, G.A., Shobha, G., Ramesh, T., Rao, K., Shobha, T.C. and Vijaykumar, T. (1995) Multicentric placebo controlled randomised double blind evaluation of a new herbal cream in vaginal infection. *Ancient Science of Life*, **14**, 212–214.

Vasanth, S., Gopal, R.H. and Rao, R.B. (1990) Plant antimalarial agent. *Journal of Scientific and Industrial Research*, **49**, 68–77.

Vashi, I.G. and Patel, H.C. (1988) Amino acid content and microbial activity of Azadirachta indica A. Juss. *Journal Institute of Chemistry*, **60** (Part II), 43–44.

Vinayagamoorthy, T. (1982) Antibacterial activities of some medicinal plants of Sri Lanka. *Ceylon Journal of Scientific and Biological Science*, **15**, 50–59.

Vohora, B.S. (1986) What is purification of blood? *Hamdard Bulletin*, **28**, 72–84.

Vohora, B.S. and Dandiya, P.C. (1992) Herbal analgesic drugs. *Fitoterapia*, **63**, 195–207.

Yadav, S.K. and Rathore, J.S. (1976) Mitotic inhibition by Melia azadirachta leaf extracts. *Proceedings National Academy of Science, India*, **46B**, 527.

Wagh, S.Y. (1988) Clinical studies in viral hepatitis. *Deerghayu International*, **4**, 17–19.

Wren, R.C. (1975) *Potter's new Cyclopaedia of Botanical Drugs and Preparations.* (reprint edition) C.W. Daniel, Essex (UK).

11. IN VETERINARY PRACTICE

TRADITIONAL USE

It is a common practice to apply neem oil alone or along with cedar wood oil externally to cattle, for any type of skin disease of any pathogenicity and even on wounds. Sometimes the animal is also made to drink the oil. It is said that neem oil aids in healing the skin, and thus gives relief to infestation. While grazing in marshy areas, the hooves of cattle often get septic. In this case, the hoof is washed with a decoction of neem and dressed with neem oil; 20–30 ml of neem oil is administered daily.

The above use of neem oil has been found useful by modern veterinarians also, and experiments have been conducted with neem oil or its compound preparations.

FOR SKIN DISEASES

Vijayan et al. (1987) prepared Oil Bordeaux from copper sulfate, quick lime and neem oil. It was administered in doses of 4 ml by intramammary infusion for 7 days. Most of the cases of mastitis recovered. Neem oil was also tried in calves, experimentally infected with the protozoa *Theileria annulata* (Srivastava et al., 1987). Antimicrobial activity was observed in a veterinary herbal antiseptic cream containing neem (Pandya et al., 1991). Neem oil was found effective in healing wounds in calves (Bhargava et al., 1991) and in camels (Purohit and Chauhan, 1992). In camels the healing process was evaluated by clinical observation, percent healing, histopathological and histochemical examination and biochemical analysis of the biopsy of specimens. The dressing material containing neem enhanced tissue repair. Anil Kumar et al. (1993) studied similar tissue-repairing activity in buffaloes.

Neem preparations have been found effective for various ectoparasitic insects. In demcodectic mange of dogs, caused by a mite, a lotion with neem soap gave very good results (Tripathy et al., 1988). A compound herbal preparation containing *Cedrus deodara*, *Azadirachta indica*, and *Embelia ribes* was tried against common poultry lice, *Mennopon gallinae* and *Liperus caponis*. This preparation caused 100 percent mortality of lice (Das et al., 1993). Another commercial preparation with the same herbal ingredients controlled canine dermatitis caused by *Demodex canis* and *Sarcoptes* spp. Hair appeared after 24–28 days and there were no symptoms of toxicity (Das and Bhatia, 1993). It was also effective in canine demodecosis in dogs with severe cutaneous lesions around the ear, neck and head, skin encrustation and pruritus due to *D. canis* (Das, 1993). In sarcoptic mange of goat, when the same preparation was sprayed, no mites (*Sarcoptes scabiei* var. *caprae*) could be found (Das et al., 1994). Adverse effects of neem preparations were observed on ticks (Williams, 1993).

For non-conventional treatments, Heath et al. (1995) applied azadirachtin to a biting louse (*Bovicola ovis*) on sheep. Azadirachtin was found quite effective as compared to the other synthetic compounds. The treatment was cost-effective in reducing louse members on the sheep for at least 40–50 days. The lack of persistence as compared with conventional insecticides was the only apparent drawback.

AS A CONTRACEPTIVE

One of the important potential uses of neem oil is its contraceptive effect. It is a spermicidal (Yao, 1993), antifertility agent and abortifacient, but could not be used in human populations because of its side effects; however, this oil can be of immense use in sterilizing stray animals by mixing in a bait for common domestic pests, rats, mice, etc. By using neem treatment we can get rid of unwanted mammals without cruelty.

REFERENCES

Anil Kumar, Sharma, V.K., Singh, H.P., Prem Prakash and Singh, S.P. (1993) Effect of some indigenous drugs on tissue repair in buffaloes. *Indian Veterinary Journal*, **70**, 42–44.

Bhargava, A.K., Lal, J., Vanamayya, P.R. and Kumar, P.N. (1991) Experimental evaluation of a few indigenous drugs as promoter of wound healing. *Vide Index Veterinarius*, 059–000005.

Das, S.S. (1993) Efficacy of Pestoban aerosol spray in treatment of canine demodecosis. *Journal of Veterinary Parasitology*, **7**, 67–69.

Das, S.S. and Bhatia, B.B. (1993) Comparative therapeutic evaluation of Ectozee aerosol spray and Betnovate-N against mite causing canine dermatitis. *Indian Journal of Indigenous Medicine*, **10**, 9–10.

Das, S.S., Bhatia, B.B. and Kumar, A. (1993) Efficacy of Pestoban-D against common poultry lice. *Indian Journal of Veterinary Research*, **2**, 25–26.

Das, S.S., Banerjee, P.S., Pandit, B.A. and Bhatia, B.B. (1994) Efficacy of a herbal compound against sarcoptic mange in goats. *Tropical Animal Health and Production*, **26**, 117–118.

Heath, A.C.G., Lampkin, N. and Jowett, J.H. (1995) Evaluation of non-conventional treatment for control of biting louse (Bouicola ovis) on sheep. *Medicinal and Veterinary Entomology*, **9**, 407–412.

Pandya, K.K. *et al.* (1991) Antimicrobial efficacy of Melicon V. A veterinary herbal antiseptic ointment. *Indian Drugs*, **28**, 255–258.

Purohit, N.K. and Chauhan, D.S. (1992) Wound healing in camels. Proceedings of the First International Camel Conference, Dubai 2nd–6th February 1992. R & W Publications, Newmarket, U.K.

Srivastava, R.V.N., Ray, D., Lal, J., Bansal, G.C. and Subramanian, G. (1987) Chemotherapeutic use of two indigenous drugs in infection of Theileria annulata in cattle. *Indian Veterinary Medicinal Journal*, **11**, 106–107.

Tripathy, S.B., Tripathy, S.N. and Das, P.K. (1988) Therapeutic efficacy of Himax (lotion) and Trichlorfon in treatment of demodectic mange of dogs. *Pashudhan*, **3**, 5.

Vijayan, R., Nair, S.P.S., Peethambaran, C.K., Balakrishnan, S., Rajan, M.R. and Oomen, S. (1987) *Kerala Journal of Veterinary Science*, **18**, 65–70.

Williams, L.A.D. (1993) Adverse effects of extracts of Artocarpus altilis Park and *Azadirachta indica* A.Juss. on the reproductive physiology of the adult female tick, Boophilus (Canest). *Invertebrate Reproduction and Development*, **23**, 159–164.

Yao, C.E. (1993) Spermicide extracted from neem. *Canopy International*, **19**, 3–4.

12. HAIRCARE AND BODYCARE PRODUCTS

The use of neem in skin diseases lead to its application on preventive aspects also. Taking a bath in a decoction of neem leaves was a ritual in some societies. The anti-inflammatory properties of neem preparations made their use more popular.

As given in Chapter 9 on Traditional uses, the neem twig is well reputed for oral hygiene, neem oil, extract or fibers have been incorporated in some of the recent toothpastes and a floss has also been prepared. Neem soap is quite popular in India and its use is also spreading in the western world. Neem extract is an important ingredient of some herbal shampoo, and neem oil is used in hair oils, body lotions, creams and mosquito repellent preparations. Neem oil is said to prevent baldness and greying of hair, and has anti-lice and anti-dandruff effects. Patents for these products have also been taken out (Sawanbori *et al.*, 1977; Latiff, 1979).

Neem has been incorporated in face packs. A typical formulation may have a very fine powder of leaves, bark and seed in clay. Milan Mehtra (1997) has given some formulations incorporating neem for face packs for oily skin, hair oil and cream for cracks on the back of the heel. In face packs, neem has been mixed with *Carica papaya* which contains papain and with liquorice. The author has suggested that a bath oil based on neem can be applied immediately after swimming to remove the last traces of chemicals and salts left on the body. Neem-based gels can possibly reduce the amount of clothes required by people living in a cold climate. The antiseptic and emollient properties of neem lotion can be useful for minor skin diseases. Neem along with Tulsi (*Ocimum sanctum*) has been incorporated in prickly heat powder and body talc.

In India, it was a common practice to apply coryllium (lamp black) along the side of the eye, particularly by young ladies as a beauty aid, to make eyes conspicuous. The common method of making lamp black was to take an earthen lamp, and put neem oil and a cotton wick in it. When ignited, the wick liberated copious smoke, from which lamp black could be collected by placing a brass cup, containing water for cooling, some distance away from the flame. The lamp black deposit was scraped from underneath the cup, and mixed with a small quantity of mustard oil to form a thick paste, called *kajal*.

This carbon black can also be used as a very safe, temporary hair dye, for concealing grey hair, by forming a very thin film on it using a hand glove.

REFERENCES

Milan Mehtra (1997) Nicely Neem. *Chemxcil Export Bulletin*, **31**, 25–28.
Latiff, A. (1979) Formulation for treating hair. British UK Patent Appli. 2000971, 24 Jan. 1979, vide *Chemical Abstract*, **92**, 11100H, 1980.
Sawanobori, H., Tanaka, H., Saito, K., Tekeuchi, Y., Shirasawa, K. and Saito, S. (1977) Melia azadirachta extracts for skin cosmetics. Japan Patent, 7728853 and 7728854.

13. TOXICOLOGY

INTRODUCTION

In Ayurveda and in Indian traditions, consumption of neem in one way or another is often prescribed and chronic patients were advised to make neem their way of life. The leaves have been fed to cattle, goats and camels as fodder. Birds have been reported to swallow the fruit. All these indicate the safety of neem to warm-blooded animals, but in recent times some of the studies have indicated that it is toxic beyond certain limits, particularly the oil. Ayurvedic preparations are usually polyherbal and neem is only one of the ingredients, used much below the toxic limits. Moreover, the herbs are processed extensively during the preparation of the product, which may reduce the toxicity further.

EARLIER REPORTS ON TOXICITY

Nadkarni (1954) mentioned that neem is a narcotic poison if used in large doses. It produces giddiness, dimness of sight, mental confusion, stupor, dilated pupils and steror. It also acted as a gastrointestinal irritant producing vomiting and purging. Nayar (1954) also included neem among poisonous seed.

The earlier toxicological studies were conducted on the major ingredient of neem bitter, the amorphous compound nimbidin, nimbidinic acid (Siddiqui and Mitra, 1945) and sodium nimbidinate (Bhide *et al.*, 1958) which were not found to be toxic.

RECENT STUDIES

While studying the traditional use of neem in malaria, Okpanyi and Ezeukwu (1981) found that some patients developed side effects, which may be due to the toxic nature of neem preparations, but Pillai and Santhakumari (1984) observed that nimbidin in doses up to 100 mg/kg did not elicit any foetal abnormality or adverse effects on rearing performance.

Qadri *et al.* (1984) tested a commercial preparation with 30 percent neem oil for subdermal toxicity in albino rats and found scaling of the epidermis and hyperkeratosis of the stratum corneum. No other adverse effect, even with doses of 200–600 mg/kg, was observed. Another commercial preparation with 3000 ppm of azadirachtin also had a very low toxicity in rats (Jacobson, 1989).

Obaseki *et al.* (1985) studied the biochemical effects after the oral administration of leaf extract and found a marked increase in the activity of 5′nucleotidase, leading to hepatobiliary toxicity, but Khatak *et al.* (1985) found no significant acute or sub-acute toxicity of various fractions up to 1.6 gm/kg. Singh *et al.* (1990) also observed that methanolic fractions from aqueous leaf extract were not toxic even up to doses of 200 mg/kg. Gandhi *et al.* (1988) noted that 5 ml/kg of neem oil was very much tolerated by animals. After a 10 ml/kg dose, mild to moderate stuporous states were observed, followed by a tremulous gait, along with signs of severe respiratory distress, followed by death in some cases. Based on the observed 24 h mortality in the various treatment groups, the LD1 and LD50 estimated by the graphic method were found to be 4.9 and 20.0 ml/kg respectively. Gross examination of the dying rat showed that all organs except the lungs looked normal. The serum value of total bilirubin and SGOT were significantly higher in the test animals as compared to the control, suggesting an early damage of liver function. Lai et al. (1990), on the other hand, observed that small amounts of neem oil given to neonates and infants on a regular basis caused toxic encephalopathy. The usual features were drowsiness, tachypnoea, and recurrent generalized seizures. Leukocytosis and metabolic acidoses was also observed.

Studies by Komolafe et al. (1988) in the case of leaf extract indicated that mixed function oxidases (MFO) are involved in the hepatobiliary toxicity, since pre-treatment with MFO inhibitors were protective. Akah et al. (1992) studied the effect of aqueous extract of neem leaves on rabbit liver using enzyme indices of hepatic dysfunction. The results indicated that a high oral dose (2328 mg/kg) of the aqueous extract may have some hepatobiliary toxic effects. Chattopadhyay et al. (1993) carried out biochemical and toxicity studies with neem leaf extract. The 24 h LD50 of the extract was 4.57 g/kg in mice. The leaf extract lowered serum cholesterol level significantly without altering serum protein, blood urea and uric acid levels.

While studying reproductive toxicity due to extract of seed in the adult cyclic female rat, Prakash et al. (1991) observed that the extract caused degenerative changes in the reproductive organs, leading to breakage and deterioration of the luminal epithelium. Total protein and glycogen contents decreased with the period of treatment. Badri Srimannarayana (1993) also concluded that a chronic administration of neem oil to adult rats for 8 days was not well tolerated and there were microscopic lesions both in the liver and the kidney. All the biochemical and histological parameters showed marked changes, indicative of the toxic effect of the oil.

The acute and sub-acute oral toxicity of azadirachtin-based pesticides has been determined by Mahboob et al. (1995) in rats. The study revealed that medium and high doses of azadirachtin caused alterations in the detoxification enzymes of various tissues, whereas low doses produced no such effects. In the case of high doses, the symptoms produced were reversible on cessation of treatment.

NEEM OIL AND REYE'S SYNDROME

Neem oil as a cause of Reye's syndrome was first observed by Sinniah and Baskaran (1981). These authors investigated the oil further to find out if the poisoning is due to oil itself or due to aflatoxin. It has been observed that neem seeds are highly contaminated with fungal flora and rich in aflatoxin, when used for oil extraction. This is mainly because neem fruit matures in the rainy season, and it has to be dried in very humid conditions. Moreover, as the oil is used mainly for making soap and other non-edible purposes, no attention is paid to cleanliness and the oil is stored and transported in used iron drums, which may cause metallic contamination or even rancidity, or bring about some chemical changes in the fatty acid composition of the oil, giving rise to toxicity. Sinniah *et al.* (1983) while pursuing the matter further, found that the aflatoxin B and C content of unrefined commercial oil varied between 40 and 100 mg/kg. It is pertinent to note here that Gandhi *et al.* (1988), who obtained oil from a reliable source for their experiments, found that oil up to 5 ml/kg was well tolerated by the animals. The toxic symptoms started after 10 ml/kg, whereas the infants studied by Sinniah and Baskaran (1981) were given only a few drops of neem oil, say a maximum of 1 ml. Considering the average weight of the child as 3 kg, the dose comes out to be 0.33 ml/kg, which is much below the toxic levels of the oil reported by Gandhi *et al.* (1988). The toxic symptoms observed by Lai *et al.* (1990) in infants are also different from those seen by Sinniah and Baskaran (1981). In another study, Sinniah *et al.* (1985) observed that intra-peritoneal injections of neem oil and seed extract may produce many of the abnormalities seen in Reye's syndrome, but there was no effect on hepatic enzymes and no evidence of cerebral oedema, which may be due to the contaminants in neem oil. Koga *et al.* (1987) noted inhibition of mitochondrial functions with neem oil and seed extract in isolated rat liver, which could be reversed with coenzyme Q.

In a review article, Nat *et al.* (1991) concluded that extracts of leaf, bark and isolated limonoids show a very low toxicity, especially when taken orally, but the seed oil is toxic. The toxic principles in the oil may be fatty acids, limonoids or an unsaturated hydroxy aldehyde. The possible mode of action of these may be due to interference with hormonal regulation, inhibition of various enzymes, interaction with receptors and alteration of membrane permeability and integrity.

Prasad *et al.* (1993) reviewed the earlier literature on the toxicology of neem and concluded that there is no mutagenic toxicity, as studied in the Ames test using *Salmonella typhimurium*.

NEEM POLLEN ALLERGY

Allergy due to neem pollens was reported by Saha and Lalyanasundram (1962). Two free amino acids were isolated from the pollens by Chanda *et al.* (1975), while Karmakar and Chatterjee (1994) isolated two allergenically active compounds AIaI and AIa IVb.

REFERENCES

Akah, P.A., Offiah, V.N. and Onuogu, E. (1992) Hepatotoxic effect of Azadirachta indica leaf extracts in rabbits. *Fitoterapia*, **63**, 311–319.

Ali, B.H. (1987) The toxicity of Azadirachta indica leaves in goats and guinea pigs. *Veterinary and Human Toxicology*, **20**, 16–19.

Badri Srimannarayana, P., Sampathraj, R. and Vanithakumari (1993) Rat toxicity study with neem oil (abstr.). *Proc. World Neem Conference*, 24–28 February, 1993, Bangalore, India.

Bhargava, K.P., Gupta, M.B., Gupta, G.P. and Mitra C.R. (1970) Anti-inflammatory activity of saponins and other natural products. *Indian Journal of Medical Research*, **58**, 724–730.

Bhide, N.K., Mehta, D.J., Attakar, M.W. and Lewis, R.A. (1958) Toxicity of sodium nimbidinate. *Indian Journal of Medical Science*, **12**, 146–148.

Chanda, S., Gangulay, P. and Mondal, S. (1975) Free amino acid composition of some allergenic pollen. *Transaction Bose Research Institute*, **38**, 3–4.

Chattopadhyay, R.R., Sarkar, S.K., Ganguly, S.K. and Basu, T.K. (1993) Biochemical and toxicity studies with Azadirachta indica leaf extract. *Journal of Ecobiology*, **5**, 233–235.

Gandhi, M., Lal, R., Sankaranaryana, A., Banerjee, C.K. and Sharma, P.L. (1988) Acute toxicity study of the oil from Azadirachta indica seed (neem oil). *Journal of Ethnopharmacology*, **23**, 39–51.

Ibrahim, I.A., Khalid, S.A., Omer, S.A. and Adam, S.E.I. (1992) On the toxicology of Azadirachta indica leaves. *Journal of Ethnopharmacology*, **35**, 267–273.

Jacobson, M. (1989) Pharmacology and toxicology of neem. In M. Jacobson (ed.) *Focus on Phytochemical Pesticides*, Vol. I. *The Neem Tree*. CRC Press Inc., Boca Raton, Florida (USA) pp. 133–135.

Karmakar, P.R. and Chatterjee, B.P. (1994) Isolation and characterisation of two IgE reactive proteins from Azadirachta indica pollen. *Molecular and Cellular Biochemistry*, **131**, 87–96.

Khatak, S.G., Gilani, S.N. and Ikram, M. (1985) Antipyretic studies on some indigenous Pakistani medicinal plants. *Journal of Ethnopharmacology*, **14**, 45–51.

Koga, Y., Yoshida, I., Kimura, A., Yoshine, M., Yamashita, F. and Sinniah, D. (1987) Inhibition of mitochondrial function by margosa oil: possible implications in the pathogenesis of Reye's syndrome. *Pediatric Research*, **22**, 184–187.

Komolafe, O.O., Anyabuike, C.P. and Obeseki, A.O. (1988) The possible role of mixed function oxidases in the hepatobiliary toxicity of Azadirchta indica. *Fitoterpia*, **59**, 109–113.

Lai, S.M., Lim, K.W. and Cheng, H.K. (1990) Margosa oil poisoning as a cause of toxic encephalopathy. *Singapore Medical Journal*, **31**, 463–465.

Mahboob, M., Siddiqui, M.K.J. and Mustafa, M. (1995) Sub acute effects of a neem pesticide on some of the detoxifying enzymes of rats. *Indian Journal of Toxicology*, **2**, 1–11.

Mohanarani, F., Nagarjan, M., Vaikundaraman, T.M. and Srinivasan, N.K. (1983) Study of poisoning in children in Tirunelveli Medical College. *Medicine and Surgery*, **23**, 35–37.

Nadkarni, K.M. (1908) *Indian Materia Medica* (reprint edition, 1954). Popular Prakashan, Bombay (India), pp. 776–784.

Nat, J.M. van der, Sluis, W.G. van der, De Silva, K.T.D. and Labadie, R.P. (1991) Ethnopharmacognostical survey of Azadirachta indica A. Juss (Meliacease). *Journal of Ethnopharmacology*, **35**, 1–24.

Nayar, S.L. (1954) Poisonous seeds of India. Part III. *Journal of Bombay Natural History*, **52**, 515–532.

Obaseki, A.O., Adeyi, O. and Anyabuika, C. (1985) Some serum enzyme level as marker of possible acute effects of the aqueous extract of Azadirachta indica on membranes *in vivo*. *Fitoterapia*, **52**, 111–115.

Okpanyi, S.N. and Ezeukwu, G.C. (1981) Anti-inflammatory and anti-pyretic activity of Azadirachta indica. *Planta Medica*, **41**, 34–39.

Pillai, N.R. and Santhakumari, G. (1984) Toxicity studies on nimbidin, a potential anti ulcer drug. *Planta Medica*, **50**, 146–148.

Prakash, A.O., Mishra, A. and Mathur, R. (1991) Studies on the reproductive toxicity due to extract of Azadirachta indica (seeds) in adult cycle female rats. *Indian Drug*, **28**, 165–169.

Prasad, H.C., Mazumdar, R. and Chakraborty, R. (1993) Research on two medicinal plants from Ayurvedic system of medicine Azadirachta indica A.Juss (syn. Melia azadirachta) and Melia azadarach Linn- their past, present and future. *Deeraghayu International*, July–Sept, 24–28.

Qadri, S.S.H., Usha, G. and Jabeen, K. (1984) Sub acute dermal toxicity of neemrich-100 (tech.) to rats. *International Pest Control*, **26**, 18–20.

Saha, J.C. and Lalyanasyundram, S. (1962) Studies on pollen allergy in Pondichery. Part I. Survey of potentially allergenic plants. *Indian Journal of Medical Research*, **50**, 881–888.

Siddiqui, S. and Mitra, C. (1945) Utilization of neem oil and its bitter constituents (nimbidin) series in pharmaceutical industry. *Journal of Scientific and Pharmaceutical Industry*, **4**, 5.

Singh, P.P., Junnarkar, A.Y., Thomas, G.P., Tripathi, R.M. and Varma, R.K. (1990) A pharamacological study of *Azadirachta indica*. *Fitoterapia*, **61**, 164–168.

Sinniah, D. and Baskaran, G. (1981) Margosa oil poisoning as a cause of Reye's syndrome. *The Lancet*, February 28, 1981, 487–488.

Sinniah, D., Schwartz, P.H., Mitchell, R.A. and Arcinue, E.L (1985). Investigation of an animal model of a Reye-like syndrome caused by margosa oil. *Pediatric Research*, **9**, 1346–1355.

Sinniah, D., Verghese, G., Baskaran, G. and Koo, S.H. (1983) Fungal flora of neem (*Azadirachta indica*) seeds and neem oil toxicity. *Malaysian Applied Biology*, **12**, 1–4.

14. NEEM IN AGRICULTURE

EARLIER REPORTS ON PESTICIDE ACTIVITIES

The activity of neem against locusts, though not well documented, has been well known to Indian farmers since very early times and some information about it is available in the earlier publications (Pruthi, 1937; Volkonsky, 1940; Pradhan *et al.*, 1962; Mitra, 1963). It was mentioned that locusts avoided feeding on leaves sprayed with crude extracts of neem and China berry. It was Robert Larson of Vikwood Botanicals, USA, who during his many business trips to India, brought to the notice of American scientific workers the property of neem against insects. This was the era when the use of synthetic pesticides was widespread, and more and more health hazards about them were coming to light, but no alternative was in sight. There was a need for safer and effective biodegradable pest control compounds with greater stability.

THE PROBLEMS CREATED BY SYNTHETIC PESTICIDES

It was seen that the continuous and indiscriminate use of synthetic chemicals for the control of insects led to the following problems:

1. Environmental pollution, as the chemicals brought about biochemical changes in the various organisms.
2. Health hazards due to high residue levels.
3. Indiscriminate destruction of insects, without any consideration of their beneficial or harmful nature.
4. Poisoning of warm-blooded animals like birds, farm animals, fish and persons coming into direct contact with these.
5. Development of resistance by insects.
6. Resurgence of certain major and minor pests, which were earlier being dominated by the insects, which were destroyed by the pesticide. With their disappearance there was less competition and new pests appeared.

PESTICIDES FROM PLANTS

Keeping the above point in view, a search for a phytopesticide started; it was seen that about 2500 plants had one or more activities against insects but only neem was found to be a highly effective, non-toxic, and environmentally friendly agent for controlling insects by acting as feeding inhibitor and growth regulator (Warthen, 1979), and it was projected as the insecticide of the future for protection against field pests (Jotwani and Srivastava, 1981). Thakur *et al.* (1981) published a bibliography on neem.

Kubo and Klocke (1982), while looking for limonoids as insect controlling agents, isolated and identified azadirachtin as an antifeedant. It was also observed that these limonoids prevented the completion of larval moulting by inhibiting the exuviae after the formation of new cuticle. These compounds did not kill the insects directly but lowered their growth rate and made them more vulnerable to other mortality factors. Jaipal et al. (1983) also noted juvenile hormone-like activities in the bark of neem, and observed that the metamorphosis of the insect was inhibited to varying degrees by these. The use of purified extract of neem was suggested for pest control. Swaminathan (1983) brought forward the potentiality of neem in pest control. Freeman and Andow (1983) described the role of neem as a tree for protection of other plants as an insect feeding deterrent. Jacobson (1986) gave details of its insecticidal activity.

With the above publications, the importance of neem became well known in the scientific world, and it became a topic of discussion at various international conferences. Schmutterer and Ascher (1986) edited the proceedings of a conference which had research papers on the pesticidal activity of neem. Saxena (1987) brought forward the use of neem as an antifeedant in pest management in the tropics and recommended quality control and standardization of its biological properties for introduction on a commercial scale. Kareem et al. (1987) meanwhile observed that with the use of neem oil mixed with custard apple (*Anona reticulata*) in rice fields, virus incidence was significantly less, and the yield of rice was higher. Singh and Singh (1988) also noted antiviral activity of leaf and bark extract of neem.

AZADIRACHTIN AS AN INSECTICIDE

Pest control aspects of neem were found to be useful in both developing and industrialized countries by Schmutterer (1988), who observed that azadirachtin and azadirachtin-containing neem extract acted as an antifeedant growth regulator and sterilant. The mode of action of azadirachtin may be due to interference with the neuroendocrine system controlling ecdysone and juvenile hormone synthesis and to inhibition of ecdysone release from the hormone-producing gland. In addition, azadirachtin causes inhibition of chitin synthesis. Azadirachtin was found to be an unstable compound, whose residual effect lasted for 4–8 days, but degradation may be hastened by ultra-violet light, rainfall and other environmental factors.

Jacobson (1989), under phytochemical pesticides, covered various aspects of this tree in the book edited by him, laying stress on its insecticidal and insect repellent properties. In this book, Ascher and Meisner (1989) discussed the effect of neem on insects attacking various crops and also on beneficial insects like the honey bee. Warthen (1989) compiled the literature on the pesticide activity of neem for the years 1979–1989. All the organisms, including arthropods, molluscs and nematodes, were covered. In a symposium on the insecticides of plant origin (Arnason et al., 1989), Saxena (1989) described insecticide from neem, and Remboldt (1989) gave an account of the structure and mode of action of azadirachtin. In another symposium, Klocke and Barnby (1989) discussed azadirchtin as an antifeedant. Powell (1989), in yet another symposium on higher plants as a source of new insecticide compounds gave a detailed account of azadirachtin. Rovesti and Deseo (1990) further discussed the potentiality of neem in pest control. Arnason and Philogene (1991), in memoirs of the entomological society of Canada, gave an account of plant-derived substances in insect control. Isman et al. (1991) studied variations in the azadirachtin content of twelve commercial samples of neem by their growth inhibition, antifeedant and moulting disruptive activity and concluded that the bioactivity of neem oil was dependent on its azadirachtin content. The possibility of a neem-based insecticide for Canada was discussed. Maramorosch (1991) reviewed the current status of research, while Remboldt and Raychaudhuri (1991) gave further details of the growth-inhibiting properties of azadirachtin. Champagne et al. (1992) described the biological activity of limonoids from neem and the other members of the *Rutales* family. Mordue and Blackwell (1993) presented an update on azadirachtin. The potential and limitations of neem pesticide were reviewed by Soon and Bottrell (1994). The authors outlined the use of neem to control pests and its effect on non-target organisms like the honey bee, earthworms, aquatic life, man and other warm-blooded animals. In a workshop (Kleeberg, 1994), the production of neem ingredients, pheromones, and their effect on phytophagus insect pests, fresh water snails and pathogenic fungi were discussed. Remboldt (1994) gave a further account of azadirachtin and its mode of action.

In India, Bambarkar (1990) presented a table showing the activity of neem against twenty species of insect pests, ten nematodes and several fungi, while Subramanyam (1990) reviewed literature on the growth-disruptive effect of neem on insects. Singh and Kataria (1991) evaluated neem against insects, nematodes and fungi. Gujar (1992) reviewed briefly the latest developments and suggested a need for standardization of formulations by using biological standards. An interesting observation was made by Gupta (1992), who isolated antifeedant microorganisms from the leaves and endosperm of neem, active only in sunlight. Tewari (1992), in his book *Monograph on Neem*, devoted a chapter to pest mangement, stressing its use in the practice of forestry. Saxena (1993) suggested it as a source of natural insecticide under integrated pest management, Nagasampagai et al. (1993) discussed bioefficacy, toxicity and compatibility of commercial products and Singh (1993) bioactivity against insect pests. Randhawa and Parmar (1993) edited the book *Neem Research and Development*. Sidhu (1995) discussed the role of neem in pest management in forestry.

The above exhaustive studies confirmed the application of neem in the fight against pests. It was found to act on eggs, larvae/nymph and adults.

Mode of Action of Azadirachtin

The various studies showed that the mode of action (Fig. 16) may be as follows:

1. Antifeedant through mouth.
   (a) Primary: it inhibits the activity of sensory receptors of mouth parts, distorts normal probing feeding and intake of food.
   (b) Ingestion of active ingredients through food leads to starvation and death.
2. Dermal action: it enters through the cuticle of the insects and inhibits chitin synthesis, thus causing desiccation and death.

3. Repellent effect: due to change in the locomotor and settling behavior of insects, in some cases mating as well as sexual communication is disrupted.
4. Growth-disruptive effect: by inhibition of the normal growth of the insect by interfering in the moulting cycle. It suppresses the activity of ecdysone so the larva does not moult, but remains at the young stage and dies.
5. Effect on survival and reproduction by oviposition deterrent action: when the female comes to an egg-laying period of her life cycle, the egg laying is prevented.
6. Effect on endocrine system: neem preparations are accumulated in the neurosecretory system and, by penetrating the blood brain barrier, are concentrated in the corpus cardiacum, resulting in reduced turnover of neurosecretory proteins.

Neem does not have an immediate knock-down effect like most of the synthetic chemicals and thus it is effective against those insects that have now become resistant to chemicals. It was also found effective against those pests that live concealed and well protected in the plant parts. Neem is not universal in its effect, which varies from insect to insect, lepidoptera being more sensitive to it, as compared to others.

RECENT STUDIES

Neem has also been found to have a nematode-suppressant activity; Siddiqui and Alam (1990), Sen and Dasgupta (1990) and Alam (1993) described this.

In 1995, quite a number of books were published on neem for its use on agricultural and domestic pests. In the book, *Neem—a user's manual*, Vijayalakshmi *et al.* (1995) gave simple methods of neem preparation from fresh materials, for control of insects by marginal farmers in India. The authors gave a brief account of the mode of action of neem products on some of the selected pests. Line drawings of the insects were given, along with their zoological names. The method of application of these on various crops has also been mentioned. Mariappan (1995) edited a book *Neem for the Management of Crop Diseases*. The book deals with the application of neem in plant diseases caused by

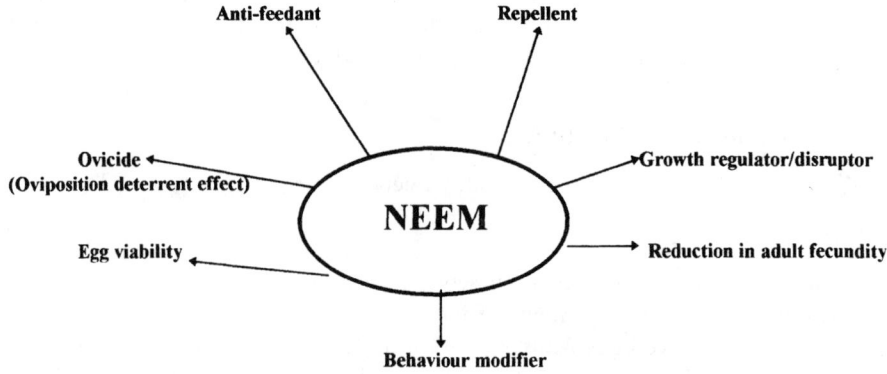

Figure 16 Diagram showing physiological effects of neem products on insects

agents other than insects, such as fungi, bacteria, viruses and nematodes in various crops of India. The results of research in applied entomology by a working group in the last decade at Giessen University are given by Schmutterer (1995). The research was carried in Latin America and in African and South Pacific countries, with very encouraging results. In some cases, by application of neem, use of synthetic pesticides could be reduced considerably, with appreciable increase in yield.

Schmutterer (1995a) edited a very exhaustive book on neem. The book carries articles on nearly all aspects from specialists who worked on neem in the last decade or so. It has detailed information on integrated pest management. The various articles deal with the use of neem pest-wise and crop-wise and also in farms and orchards. Singh et al. (1995) edited the book *Neem and Environment*. The book contains articles presented at the world conference on neem held in 1993.

REFERENCES

Alam, M.M. (1993) Bioactivity against phytonematodes. In N.S. Randhawa and B.S. Parmar (eds.) *Neem Research and Development*. Pub No. 3. Society of Pesticides Science, India, pp. 123–143.

Arnason, J.T. and Philogene, B.J.R. (1991) Symposium on the role of plant derived substance for insect control. *Memoirs of the Entomological Society of Canada*, No. **159**, 47.

Arnason, J.T., Philogene, B.J.R. and Morand, P. (eds.) (1989) Insecticides of plant origin. *American Chemical Society Symposium*. Series **387**, Washington DC.

Ascher, K.R.S. and Meisner, J. (1989) The effects of neem on insects affecting man and animal. Focus on phytochemical pesticides. In M. Jacobson (ed.), *Focus on Phytochemical Pesticide*. Vol. I *The Neem Tree*. CRC Press, Florida, USA.

Bambarkar, S. (1990) Neem—a vast potential for agrochemicals. *Pesticide, Bombay*, India, **24**, 36–38.

Champagne, D.E., Koul, O., Isman, M.B., Scudder, G.G.E. and Towers, G.H.N. (1992) Biological activity of limonoids from the Rutales. *Phytochemistry*, **31**, 377–394.

Freeman, A.B. and Andow, D.A. (1983) Plants protecting plants: the use of insect feeding deterrent. *Scientific Horticulture*, **34**, 48–53.

Gujar, G.T. (1992) Neem (*Azadirachta indica* A. Juss)—a natural insecticide: status and need for bioassay standards. *Pesticide Research Journal*, **4**, 69–79.

Gupta, M. (1992) Micro-organism in neem. *Indian Journal of Entomology*, **54**, 359–363.

Isman, M.B., Koul, O., Arnason, J.T., Stewart, J. and Sallocum, G.S. (1991) Developing a neem based insecticide for Canada. *Memoirs of the Entomological Society of Canada*, No. **159**, 39–47.

Jacobson, M. (1983) Insecticides, insect repellants and attractants from arid/semiarid land plants. Plant—the potential for extracting protein, medicines and other useful chemicals. *Workshop proceedings OTA-BP-F23*, 138–146. US Congress Office of Technology Assessment, Washington DC, USA.

Jacobson, M. (1986) The neem tree: natural resistance par excellence. In M.B. Green and P.A. Hedin (eds.) Insecticides of the plant origin. Natural Resistance of Plants to Pests. *American Chemical Society Ser.*, **296**, 220–232.

Jacobson, M. (ed.) (1989) *Focus on Phytochemical pesticides*. Vol. I. *The neem tree*. C.R.C Press Inc., Florida, USA.

Jaipal, S., Zile, S. and Chuhan, R. (1983) Juvenile hormone like activity of some common Indian plants. *Indian Journal of Agricultural Sciences*, **53**, 730–733.

Jotwani, M.G. and Srivastava, K.P. (1981) Neem insecticide of future. II. Protection against field pests. *Pesticide, Bombay*, **15**, 40–47.

Kareem, A.A., Saxena, R.C. and Justo, H.D. Jr. (1987) Cost comparison of neem oil as an insecticide against rice tungro virus (RTV). *International Rice Research Newsletter*, **12**, 28–29.

Kleeberg, H. (1994) Practice oriented results on use and production of neem ingredients and pheromones. *Proceedings of the 3rd Workshop*, Wetzlar, Novemeber 22–25, 1993. Druck and Graphic, Giessen, Germany.

Klocke, J.A. and Barnby, M.A. (1989) Plant allelochemical as source and models of insect control agent. Phytochemical ecology: allelochemicals, mycotoxins, and insect pheromones, and allomones. In C.H. Chou and G.R. Waller (eds.) *Proc. of a Symposium held in Taipei and Kengting, Taiwan*.

Kubo, I. and Klocke, J.A. (1982) Limonoids as insect control agent. Les mediateurs chimiques agissant sur le comportement des insectes. *Symposium International*, Versailles, 16–20 Nov., 1981.

Mariappan, V. (ed.) (1995) *Neem for the Management of Crop Diseases*. Associated Publishing Co., New Delhi, India.

Maramorosch, K. (1991) Current status of neem pesticide and by products *Recent Advances in Medicinal Plant Spice Crop*, No. **1**. Today and Tomorrow Printers & Publishers, New Delhi, India.

Mitra, C.R. (1963) *Neem*. Central Oil Seed Committee, Hyderabad.

Mordue, A.J. and Blackwell, A. (1993) Azadirachtin: an update. *Journal of Insect Physiology*, **39**, 903–924.

Nagasampagai, B.A., Kulkarni, M.M., Rojatkar, S.R. and Ayangar, N.R. (1993) Utilisation of neem in pest management. *Indian Society of Tobacco Science*, Rajahmundry, India.

Powell, R.G. (1989) Higher plants as source of new insecticidal compounds. *Pesticide-Science*, **27**, 228–229.

Pradhan, S., Jotwani, M.G. and Rai, B.K.(1962) The neem seed deterrent to locusts. *Indian Farming*, **12**, 7–11.

Pruthi, H.S. (1937) Report of the Imperial entomologist 1935–36. *Scientific Report of the Agricultural Research Institute, New Delhi*.

Randhawa, N.S. and Parmar, B.S. (eds.) (1993) *Neem Research and Development*. Pub No. **3**, Society of Pesticide Science, India.

Remboldt, H. (1994) Azadirachtin—a botanical insect growth inhibitor and its relation to biosemiotics. *Proceedings of the Indian National Science Academy*, Part B, *Biological Sciences*, **60**, 471–486.

Remboldt, H. (1989) Azadirachtins: their structure and mode of action. In: J.T. Arnason, B.J.R, Philogene and P. Morand (eds.) Insecticides of plant origin. *American Chemical Society Symposium*. Series **387**, Washington DC, pp. 150–163.

Remboldt, H. and Raychaudhuri, S.P. (1991) The azadirachtins—highly active insect growth inhibitors. *Recent Advances in Medicinal, Aromatic and Spice Crops*. Vol. I. Today and Tomorrow Printers and Publishers, New Delhi, India, pp. 31–37.

Rovesti, L. and Deseo, K.V. (1990) *Azadirachta indica* A. Juss (neem) and its potential in the control of insects. *Informatore Fitopatalogico*, **40**, 27–32.

Saxena, R.C. (1987) Antifeedants in tropical pest management. Insect Science and its application. *Recent Advances in Research on Tropical Entomology*, Nairobi, Kenya, 31 August to 5 September, 1986, pp. 731–736.

Saxena, R.C. (1989) Insecticide from Neem. In J.T. Arnason, B.J.R. Philogene and P. Morand (eds.) Insecticides of Plant Origin. *American Chemical Society Series*, No. **3879**, 110–135.

Saxena, R.C. (1993) Neem as a source of natural insecticide—an update. *Botanical Pesticides in Integrated Pest Management*, Indian Society of Tobacco Science, Rajhamundry, India.

Schmutterer, H. (1988) Potential of azadirachtin containing pesticides for integrated pest control in developing and industrialized countries. *Journal of Insect Physiology*, **34**, 713–719.

Schmutterer, H. (1995) Result of research in applied entomology in the tropics, obtained by working group at Giessen University during the last ten years. *Mitteilungen der Deutschen Gesellschaft fur Allegemeine und Angewandte Entomologie*, **10**, 277–286.

Schmutterer, H. (1995a) *Neem Tree, source of unique natural products for integrated pest management and medicinal, industrial and other purpose*. VCH Verlag, Weinhim, Germany.

Schmutterer, H. and Ascher, K.R.S. (eds.) (1986) Natural pesticides from the neem tree (*Azadirachta indica* A. Juss.) and other tropical plants. *Proc. 3rd Neem Conf.*, Nairobi, Kenya, 10–15 July, 1986.

Sen, K. and Dasgupta, M.K. (1990) Control of root nematodes (Meloidogyne spp.) with organic amendments and soil nematadocides on some successive crops. *Horticultural Journal*, **2**, 48–54.

Sidhu, D.S. (1995) Neem in agroforestry as a source of plant derived chemicals for pest management. *Indian Forestery*, **121**, 1011–1021.

Singh, A.K. and Singh, M. (1988) Antiviral and physical properties of the extracts of *Azadirachta indica* L. *Indian Journal of Virology*, **4**, 76–81.

Singh, R.P. (1993) Bioactivity against insect pests. In N.S. Randhawa and B.S. Parmar (eds.) *Neem Research and development*. Society of Pesticide Sciences, India. Pub No. **3**, 109–110.

Singh, R.P., Chari, M.S., Raheja, A.K. and Kraus, W. (1995) *Neem and Environment*. Vol. I, II. International Book Distributors, Dehradun, India.

Singh, R.P. and Kataria, P.K. (1991) Insects, nematodes and fungi evaluated with neem (*Azadirachta indica* A. Juss) in India. *Neem Newsletter*, **8**, 3–10.

Siddiqui, M.A. and Alam, M.M. (1990) Sawdust as soil amendments for control of nematodes infesting some vegetables. *Biological Wastes*, **33**, 123–129.

Soon, L.G. and Bottrell, D.G. (1994) *Neem pesticides in rice: potential and limitations*. International Rice Research Institute, Manila, Philippines.

Subramanyam, B. (1990) Azadirachtin—a naturally occurring insect growth regulator. *Proceedings of the Indian Academy of Science, Animal Science*, **99**, 277–288.

Swaminathan, M.S. (1983) *Neem in Agriculture*. Indian Agriculture Research Institute, New Delhi, India.

Thakur, R.S., Singh, S.B. and Goswami, A. (1981) *Azadirchta indica* A. Juss.—A review. *Current Research in Medicinal and Aromatic Plants*, **3**, 135–140.

Tewari, D.N. (1992) *Monograph on neem*. International Book Distributors, Dehradun, India.

Vijayalakshmi, K., Gaur, H.S. and Goswami, B.K. (1995) Neem for the control of plant parasitic nematodes. *Neem Newsletter*, **2**, 35–42.

Volkonsky, M. (1940) The acridfuge action of extracts from the leaves of Melia azaderach. *Indian Forester*, **66**, 53–56.

Warthen, J.D. Jr. (1979) *Azadirachta indica: a source of insect feeding inhibitor and growth regulators*. US Department, Agriculture Review Manual, ARM-NE **4**, 1–21.

Warthen, J.D. Jr. (1989) Neem (*Azadirachta indica* A. Juss): organisms affected and reference list update. *Proc. of the Entomological society of Washington*, **91**, 367–388.

15. NEEM SEED CAKE AS A MANURE AND NITRIFICATION INHIBITOR

AS A MANURE

The use of seed cake, obtained after the extraction of oil, particularly that of non-edible seed, as a manure is well known in agriculture. The use of neem cake has also been recommended for this purpose. Keeping in view the importance of this in agriculture, the Indian Standards Institute (Anonymous, 1977) has given specification No. 8558 for neem cake, for manuring, which is as follows:

- Maximum moisture (percent by mass) 10%
- Maximum water soluble organic nitrogen on moisture free basis (percent by mass) 2.5%
- Maximum total ash (percent by mass) 13%
- Maximum acid insoluble ash (percent by mass) moisture free basis 5.0%

Neem seed cake not only provides nutrition to the plant, but controls soil-borne pests, particularly nematodes; it also acts as a nitrification inhibitor, helps respiratory activity, increases the population of earthworms and produces organic acids, which help in removing the alkalinity of the soil (Korah and Shingte, 1968).

AS A NITRIFICATION INHIBITOR

When a nitrogenous fertilizer is applied to the soil, about half the nitrogen escapes into the air by the activities of nitrifying bacteria, such as *Nitrosomonas* and *Nitrobacter.* If a bacteriostatic like sulfur is used, it suppresses the growth of these bacteria and acts as a nitrification inhibitor (Rajendra Prasad *et al.*, 1971). The same results were obtained by Sahrawat and Parmar (1975) from an alcoholic extract of neem. Neem-coated urea increased the apparent recovery of nitrogen to 48.6 percent, which was 29.3 percent in the case of untreated urea (Ketkar, 1976). In a field experiment, 160 kg urea with 40 kg neem cake per hectare produced 6 quintals more rice, as compared to the field where only 200 kg urea was applied. Ramababu *et al.* (1983) got similar results.

Watanabe *et al.* (1981) studied the effect of neem cake on the nitrogen fixation activity of blue green algae in a flooded rice field. Slangen and Kerkhoff (1984) presented a review of literature on nitrification inhibitors in agriculture and horticulture. Uma Singh and Gurumurti (1984) suggested neem cake as a potential fertilizer. In a study it was seen that after ten months, the root length, number of branches and dry weight of the plants increased as compared to the control (Uma Singh *et al.* 1986).

Naidu and John (1984) found that fermented neem cake, when used with other oil seed cakes, inhibited the fungal rice pathogens *Rhizocotonia solanii* and *Sclerotium oryza* (*Magnaporthe salvinii*). Korah and Shingte (1988) presented data to show that nitrogen mineralization was greatest with neem as compared to other non-edible oil seed cakes. Nair and Sharma (1988) noted the difference in the performance of ordinary and neem-coated urea. The persistence of neem kernel powder and neem oil on phorate granules in the soil was studied by Dethe and Babtiwale (1989), and it was found that degradation was faster in unsterilized soil as compared to sterilized soil. Ali and El-Sanousi (1989) tested the sensitivity of different aerobic bacterial species to azadirachtin, while Turker *et al.* (1989) studied the effect of different nitrificaton inhibitors including neem on protein, non-protein, ammonical and nitrate content in the soil. Rajinder Prasad *et al.* (1990) also carried out this type of study.

Thampatti *et al.* (1992) applied muriate of potassium with urea neem cake blend and found that the exchangeable potassium was highest in all soil layers, as compared to the control.

Many studies, including the one by Ameta and Singh (1990), confirmed an increase in yield when urea coated with neem powder was applied to the field, but Das and Mukherjee (1990) noticed the adverse effects of neem cake on beneficial organisms present in the soil. Mukherjee *et al.* (1991) confirmed this and observed a lesser number of bacteria, fungi and actinomycetes. It was also observed that neem seed cake caused an accumulation of ammoniated nitrogen in the soil during inhibition of nitrification of urea and an increase in pH, which gives rise to several problems, including damage to young plants due to nitrogen toxicity.

The denitrification property of neem cake has also not been properly understood so far. It may be due to sulfur compounds which have a well-known antibacterial effect, or the lipids associated with the cake may be responsible for inhibition of bacterial growth. The activity may be due to tannins which inhibit the urease enzyme in the soil. Devakumar and Riar (1993) considered it due to bioregulators like meliacins, epinimbin, salin and azadirachtin.

For large-scale production of neem-coated urea, Vyas *et al.* (1991) developed a neem extract. This was an improvement on the earlier method, in which coal tar dissolved in kerosene oil was sprayed on urea as an adhesive, followed by a coating of fine powder of neem cake on each granule of urea. Suri (1995) suggested a commercially viable process in which an emulsion of neem oil was coated on urea during the production stage.

REFERENCES

Ali, B.H. and El-Sanousi, S. (1989) Sensitivity of different aerobic bacterial species to azadirachtin. *Journal of Arid Environment*, **17**, 117.

Ameta, G.S. and Singh, H.G. (1990) Comparative efficacy of neem cake coated prilled urea and splitting N-application in rice production. *International Journal of Agriculture Tropical*, **8**, 189–192.

Anonymous (1977) *Requirement of neem cake for manuring*. Indian Standards Institute, Specification No. 8558.

Das, A.C. and Mukherjee, D. (1990) Microbiological changes during decomposition of wheat straw and neem cake in soil. *Environment and Ecology*, **8**, 1012–1015.

Dethe, M.D. and Babtiwale, H.P. (1989) Persistence of neem products coated phorate granules in soil. *Journal of Ecobiology*, **1**, 265–267.

Devakumar, C. and Riar, S.S. (1993) Identification of bioregulators in neem: Results of ten year studies. (Abstr.). *Proc. World Neem Conference*, 24–28 February 1993, Bangalore, India.

Ketkar, C.M. (1976) *Utilization of Neem and its by products*. Khadi and Village Industries Commission, Bombay, India.

Korah, P.A. and Shingte, A.K. (1968) On the effect of non-edible oil cakes on the respiratory activity of soil. *Agricultural Research Journal of Kerala, India*, **6**, 95–97.

Korah, P.A. and Shingte, A.K. (1988) The mineralization of nitrogen from non-edible oil cakes in medium black calcareous and laterite soils. *Agricultural Research Journal of Kerala, India*, **26**, 215–226.

Naidu, V.D. and John, V.T. (1984) Anti-fungal activity of degraded oil cake extracts. *Journal of Research, Assam Agricultural University*, **5**, 209–210.

Nair, K.P.P. and Sharma, P.B. (1988) Comparative effectiveness of ordinary and coated urea and nitrification inhibitor treated urea as a source of nitrogen for maize. *Experimental Agriculture*, **24**, 477–479.

Mukherjee, D., Mitra, S. and Das, A.C. (1991) Effect of oil cakes on changes in carbon, nitrogen and microbial population in soil. *Journal of the Indian Society of Soil Science*, **39**, 457–462.

Prasad, Rajinder, Rajale, G.B. and Lkahdivi, B.A. (1971) Nitrification retarder and slow release nitrogen fertilizers. *Advance in Agronomy*, **23**, 337–383.

Rajendra Prasad, Sharma, S.N., Surendra Singh and Mangal Prasad (1990) Nitrogen Management. Soil fertility and fertilizer use. Vol. IV. In Virendra Kumar G.C., Shotriya and S.V. Kaore (eds.) *Nutrient Management and Supply System for Sustaining Agriculture in 1990*. IFFCO, New Delhi, India, pp. 41–51.

Ramababu, P., Reddy, S.N. and Pillai, K.G. (1983) Effect of super granules and coated urea on productivity of rice. *Agricultural Science Digest*, **3**, 194–202.

Sahrawat, K.L. and Parmar, B.S. (1975) Alcoholic extract of neem *Azadirachta indica* seed as nitrification inhibitor. *Journal Indian Society of Soil Science*, **23**, 131–134.

Slangen, J.H.G. and Kerkhoff, P. (1984) Nitrification inhibitor in agriculture and horticulture a literature review. *Fertilizer Review*, **5**, 1–76.

Suri, I.K. (1995) Coating of urea with neem. *Fertilizer News*, **40**, 55–59.

Thampatti, K.K.C., Padmaja, P. and Manorama (1992) Use of nitrification inhibitors and N-K interaction in laterite soil. *Journal of Potassium Research*, **8**, 200–209.

Turker, O.R., Pathak, A. and Dikshit, P.R. (1989) Protein and non-protein nitrogen content in wheat and ammonical and nitrate nitrogen content in soil, as influenced by urea treated with different nitrification inhibitors. *Bhartiya Krishi Anusandhan Patrika*, **4**, 75–83.

Uma Singh and Gurumurti, K. (1984) Oil cakes from oil seeds of forest origin and their potential on fertilizer. *Indian Journal of Forestry*, **7**, 12–18.

Uma Singh, Rawat, P.S. and Purohit, C.K. (1986) Dry matter production in response to application of tree seed oil cake as fertilizer in Leucaena leucocephala. *Indian Journal of Forestry*, **10**, 214–216.

Vyas, B.N., Godrej, N.B. and Mistry, K.B. (1991) Development and evaluation of neem extract as a coating for urea fertilizer. *Fertilizer News*, **36**, 23–25.

Watanabe, I., Subudhi, B.P.R. and Aziz, I. (1981) Effect of neem cake on the population and nitrogen fixing activity of blue green algae in flooded soil. *Current Science*, **50**, 937–939.

16. POULTRY AND CATTLE FEED

In developing countries particularly, there is competition between man and other domestic animals for conventional food, leading to malnutrition for all. If regular cereals and legumes consumed by man are replaced partly by other food items in the diet of chicken, cattle, pig, etc. (Punj, 1988), it will release pressure on the food directly as well as indirectly. If there is less consumption of cereals by animals, more food will be available to the human population, and man will be saving cereals/legumes by eating a non-vegetarian diet.

Neem leaves are consumed by camels, goats and in drought by cattle also (Patel and Patel, 1957; Hentgen, 1985). Shukla and Desai (1988) suggested neem as a source of cattle feed. The seed is rich both in fatty oil and in protein, but it could not be used unprocessed for edible purposes on account of the deleterious effects it has on animals and birds because of the bitter principles contained in it. For the last fifty years, research has been conducted, and various feed trials on animals have been undertaken to utilize these seed as feed. An account of these activities is given here.

ANIMAL FEED

The Leaves

These are fed particularly to camels, which appear to relish them. Goats may also eat them. In Andhra Pradesh (India), leaves are fed to milk-yielding animals to increase the yield of milk after parturition.

Singh (1982) included neem among the fodder trees of India. (The neem tree was introduced in arid zones or on degraded soil on a large scale and the use of leaves as a forage was advocated). As a multi-purpose tree, the leaves were recommended as suitable for browsing and for incorporation into a concentrated ration after drying. It was reported that dry neem leaves have better nutritional quality than sorghum, which is the main dry season fodder (Zech and Weinstable, 1983; Webb, 1988). The leaves are said to be palatable to cattle and buffaloes; they are a good source of protein (15 percent) and carotene 185 µg/g, and contain most minerals except zinc. They have lower fiber content. Fallen leaves can also be fed but they are less palatable, containing about 8 percent protein and 12.71 percent ash. Conventional digestible trials by Patel and Shukla (1962) found that neem leaves contain 6.2 percent digestible protein and 52.5 percent total digestible nutrients. It was reported by Jayal (1963) that the digestible coefficient of neem leaves was 51.96 percent, for protein 58.45 percent and for fiber 22.33 percent.

The approximate composition and nutritive value of neem leaves as given by Ranjhan (1980) are organic matter 92.25 percent, crude protein 16.12 percent, ether extract 3.40 percent, crude fiber 20.69 percent, nitrogen-free extract 52.06 percent, total ash 7.73 percent, digestible energy 2394 Kcal/kg and metabolizable energy 1926 Kcal/kg.

Murugan and Kathaperumal (1987) and Murgan *et al.* (1987) determined the macro and micro mineral content of leaves to see if these can be used as animal fodder.

When animals are fed with cereal and fodder, the danger of a negative nitrogen and mineral balance exists; the feeding of neem leaves help to alleviate these.

It is the fodder of choice during dry periods and drought. In the west Indian state of Gujrat, during a famine 15–20 kg of neem leaves were fed to cattle and buffaloes daily for their survival (Ketkar, 1976).

The above use does not appear to be universally acceptable, as Singh and Pathak (1981) and Murugan and Kathaperumal (1987) observed that sheep fed on leaves or fruits, dry or meal or silage, lost body weight. Ali (1987), when he administered an aqueous suspension of dry leaves at a dose of 50 or 200 g/kg to goats over a period of up to 8 weeks, saw a progressive decrease in body weight, together with weakness and lack of appetite. Higher doses even produced tremors and ataxia, with drastic histopathological changes in various organs. In guinea pigs the toxic symptoms were less severe. Ibrahim *et al.* (1992) fed chicks with a diet containing 2–5 percent leaves. A decrease in body weight gain and efficiency of feed utilization was observed, with yellow discoloration of legs and combs. Many histopathological changes took place in the body.

Neem Oil

Because of bitters and odoriferous principles, the oil was not considered suitable for animal consumption. It was found to be toxic though its fatty oil composition is very near to other edible oils. Rukmani (1987) and Rukmani *et al.* (1991), to evaluate the nutritional value of the oil, extracted the seed with aqueous ethanol followed by hexane, and got oil free from bitters, odouriferous principles, coloring matter and free fatty acids. A toxicity study on neem was also carried out by Rukmini (1987), who found it to be a safe source of edible oil to be used as such or blended or hydrogenated. Reddy *et al.* (1988) concluded that oil treated with 10 percent potassium hydroxide (KOH) was detoxified, as was evident when fed to chicks. The oil, prepared by treating it with 5 percent KOH, without heat, was not completely utilized by chicks, but when given in a phased manner, it was comparable with the control diet. Vasishtha *et al.* (1992) studied solvent-extracted neem oil. Chinnasammy *et al.* (1993) carried out further studies on reproductive toxicology for three generations. There was no change in the various organs of the body or mutagenicity as compared to the control. The oil was found to be fit even for human consumption. The fatty oil composition of neem oil reported was myristic acid 0.2 to 2.6 percent, palmitic acid 13.6 to 16.2 percent, stearic acid 14.4 to 24.1 percent, oleic acid 49.1 to 61.9 percent, linoleic acid 2.3 to 15.8 percent, and archidonic acid.

Full Fat Seed Meal

Chicks, when fed a diet containing 2, 5, 10 percent ripe fruit of basic diet, from 7 to 35 days, showed a decrease in body weight gain, efficiency of feed utilization and hepatonephropathy (Ibrahim *et al.*, 1992a). Full fat seed meal at 0, 25, 75 and 100 g/kg was studied in broilers and rabbits at 0, 200, 300 g/kg for 8 weeks. In chicks, there was a significant negative relation between neem meal, weight gain and feed conversion efficiency. The rabbits, when given up to 100 g/kg of neem preparations, were superior to the control, but not at higher concentrations (Salawu et al., 1994).

Seed Cake

The easy availability of seed cake in large quantity at affordable price and high nutritive value, came to the notice of scientific workers a long time ago.

According to Ranjhan (1980), neem cake contained 82.40 percent organic matter, of which crude protein was 17.03 percent, ether extract 1.02 percent, crude fiber 42.12 percent, nitrogen-free extract 22.13 percent and total ash 17.60 percent. The percentage of amino acid profile of neem cake was aspartic acid 1.31, threonine 0.50, serine 0.38, glutamic acid 2.40, proline 0.84, glycine 1.08, cysteine 1.73, valine 0.76, methionine 0.70, isoleucine 0.60, leucine 0.95, tyrosine 0.26, phenylalanine 0.80, histidine 0.21, lysine 0.28 and arginine 0.57. Maitra and Duttagupta (1982) gave an analytical report of neem seed cake as follows: crude protein 21.9 percent, ether extract 8.6 percent, nitrogen-free extractives 44.2 percent and crude fibers 11.6 percent. It is very rich in organic minerals. In commercial cattle feed, inorganic minerals are usually added; cattle have to convert them into organic compounds and hence the availability of minerals is only up to 40 percent, the rest being lost in the process of digestion. Neem cake helps in mineral nutrition, as it is particularly rich in Ca, P, Fe, Cu and Zn. Rao (1987) carried out nutritional trials with debittered and defatted oil cake.

The chemical analysis by Reddy et al. (1988a) to establish its suitability for animal feed indicated that it compared well with peanut cake. It contained less crude protein, more fiber, fat-and nitrogen-free extractives, calcium and phosphorous. Carbohydrate content was 19.28 to 22.27 percent. True metabolizable energy value for decorticated expeller-processed neem cake was 2.959 to 2.973 Kcal/g and for undecorticated expeller-produced neem cake it was 2.279 to 2.823 Kcal/g with gross protein value 55 to 59 percent.

The only negative point with neem cake was the taste and odor, which are not palatable to animals. For removing the bitter principles, various methods were tried earlier; some of these are:

1. Treating with various organic solvents like petroleum, alcohol, etc. (Lehri et al., 1987). One of the methods suggested was to dry the seed properly, wash them with hexane to recover all the oil, and then to extract with ethanol.
2. Treating with various concentrations of acid and alkali.
3. Repeatedly washing with water until the bitter principles are removed.
4. Heating the neem cake to 150°C.

FEEDING TRIALS WITH ANIMALS

Cattle and Buffalo

Earlier work on the nutritive value of neem cake and its digestability by buffaloes was done by Arora et al. (1975) and Bedi et al. (1975). Pyne et al. (1979) studied the composition of the milk of lactating buffaloes which were given neem cake in the feed. Maitra and Duttagupta (1982) heated untreated neem cake to 150°C. Heat treatment did not affect nutrient digestibility significantly. Maitra *et al.* (1982) also observed that the net effect of energy utilization was higher for a concentration of 10 to 15 percent, as compared to a concentration of 20 percent. Nath *et al.* (1989), when they treated neem seed kernels with water and dried them, found that these were palatable to cattle. The animals could digest them well and they did not affect the physiology or even some blood constituents. Agarwal *et al.* (1987) also observed that water-washed neem kernel feed had higher efficiency of utilization of digestible protein, as indicated by the greater nitrogen balance (Singhal and Mudgil, 1983).

In field trials, various efforts were made to make neem cake palatable to animals, by adding molasses and starch. There was some acceptability when double the amount of maize was added, but in these cases the protein metabolism was adversely affected. In buffalo, the addition of 10, 15 and 20 percent neem cake did not have any adverse effect on the milk composition or the health of the animals. Both red and white blood cells and hemoglobin were higher in the blood but serum protein was lower. The serum protein was decreased by increasing the neem seed cake in ration. Some factors affected hematopoiesis (blood formation). In the liver, it did not affect SGOT and SGPT. Calcium and phosphorus in the blood was also not affected (Gangopadhyay *et al.* 1979, 1981).

The feeding of water-washed neem kernel cake to cows at 400 g/kg did not a affect performance, blood constitution or reproductive ability (Nath *et al.*, 1983, 1989). A higher nitrogen balance was caused by reduced excretion of nitrogen in urine and a decrease in blood urea nitrogen. Kumar *et al.* (1990, 1992), after the study, concluded that neem seed cake can replace up to 30 percent dairy concentrate, consisting of proteins, fats, minerals, etc. It did not affect nutrient digestibility or milk yield. In another study, concentrate was replaced by various proportions of neem cake, *Pennisetum purpureum* and mixed grass hay. The above feed did not affect milk yield or its composition. But when peanut cake was fully replaced by neem cake by Garg and Nath (1990), it depressed nutrient digestibility and also the nitrogen, calcium and phosphorus balance. When the neem seed kernel was fed with a balanced diet after soaking in a solution of sodium hydroxide and washing with water, there were no clinical signs of ill health and vital organs showed no pathological changes (Katiyar *et al.*, 1993). It was concluded that alkali treatment totally detoxified the neem cake. Water-washed neem kernel powder, when included in feed by Mahendra *et al.* (1995) for four months, did not have any adverse effect on the daily milk yield, but lowered the feed cost considerably.

Pig

A diet which contained 5 percent neem cake decreased the value of feed efficiency (Thomas and Prasad, 1983), but when neem cake was replaced by water-washed neem kernel cake (40 percent crude protein), pigs grew faster and utilized the feed more efficiently with higher nitrogen retention. The feed cost was reduced by 11 percent (Sastry and Agarwal, 1992).

Poultry

Toxicity in the water-extracted neem cake was found by Singh *et al.* (1985). Expeller processed neem cake decreased the weight and feeding efficiency of birds (Reddy and Rao, 1988), as was the case with solvent-extracted neem cake, but when neem cake treated with acid or alkali was fed to birds, it improved growth which was comparable to the control (Reddy and Rao, 1988a, 1988b; Reddy *et al.*, 1988). Chakravarty and Prasad (1991) studied the effect of neem leaf extract and other neem extracts on the performance of broiler chicks.

Rabbit

Fajinmi *et al.* (1990) studied neem seed in the diet of rabbits with encouraging results.

Rat

In feeding trials with rats also, the growth-suppressant compound was found (Vijjan and Parihar, 1983) to be water soluble and could be removed by washing neem cake (Vijjan, 1983). Prakash *et al.* (1991) observed reproductive toxicity in the case of adult females. Garg *et al.* (1991) tested this compound in males and observed a decline in growth rate, food consumption and testes weight.

Sheep and Goat

With feeding trials, weight loss and toxicity symptoms were evident in lambs. After 15 days there was severe gingivitis and sloughing of mucus membrane with foaming discharge from the mouth, after 25 days stomatitis, gastro-enteritis, and diarrhea were followed by death (Vijjan *et al.*, 1982), but when a mixture containing 17 and 34 parts of neem was fed to sheep the symptoms were less severe. Gupta and Bhaid (1981) and Ramu *et al.* (1994) fed 10, 20, 30 percent water-washed neem seed cake to sheep and goats. They digested it well. These authors carried out further studies (Ramu *et al.*, 1994a) and concluded that a complete ration can be formulated by incorporating 30 percent water-washed neem kernel seed. Evaluation of cooked meat did not reveal any bitter taste up to 20 percent when water-washed neem cake was added in the diet of sheep and goats in place of deoiled rice bran (Reddy *et al.*, 1994). In another study, Verma (1995) and Verma *et al.* (1996) also concluded that water-washed neem kernel can be incorporated in the diet of growing goats up to 25 percent without any deleterious effects on nutrient utilization and metabolism. This type of diet did not have any adverse effects on the weight of the goats, and sensory evaluation of cooked meat did not reveal any bitter taste.

REFERENCES

Agarwal, D.K., Garg, A.K and Nath, K. (1987) The use of water washed neem (*Azadirachta indica*) seed kernel cake in the feeding of buffalo calves. *Journal of Agricultural Science*, UK, **108**, 497–499.

Ali, B.H. (1987) The toxicity of *Azadirachta indica* leave in goats and guinea pigs. *Veterinary and Human Toxicology*, **20**, 16–19.

Arora, S.P., Singhal, S.P. and Kudri, R.S. (1975) Nutritive value of neem and cake (*Mellia indica*). *Indian Veterinary Journal*, **52**, 867–870.

Bedi, S.P.S., Ujjain, U.K. and Ranjhan, S.K. (1975) Utilization of neem (*Azadirachta indica*) seed cake and its influence on nutrient digestibility in buffaloes. *Indian Journal of Dairy Science*, **28**, 104–107.

Chakravarty, A. and Prasad, J. (1991) Study on the effect of neem leaf extract and neem cake extract on the performance of broiler chicks. *Poultry Adviser*, **24**, 37–38.

Chinnasammy, N., Harishankar, N., Kumar, P.U. and Rukmini, C. (1993) Toxicological studies on debitterized neem oil (*Azadirachta indica*). *Food and Chemical Toxicology*, **31**, 297–301.

Fajinmi, A.O., Adedeji, S.K., Hassan, W.A. and Babatunde, G.M. (1990) Inclusion of non-conventional feed stuffs in rabbit concentrate ration—a case study of neem (*Azadirachta indica*) seeds. *Journal of Applied Rabbit Research*, **13**, 125–126.

Garg, A.K., Agarwal, D.K., Singh, S.D. and Nath, K. (1991) Growth and spermatogenesis in rats fed solvent extracted water washed and untreated neem (*Azadirachta indica*) seed kernel cake. *International Journal of Animal Sciences*, **7**, 223–235.

Garg, A.K. and Nath, K. (1990) Solvent extracted neem (*Azadirachta indica A. Juss*) seed cake in the ration of growing cross breed calves. *Indian Journal of Animal Nutrition*, **7**, 199–202.

Gangopadhyay, P., Maitra, D.N. and Pyne, A.K. (1979) Studies on the blood constituents with the use of neem seed expeller cake (*Azadirachta indica*) in lactating Murrah buffaloes. *Indian Veterinary Journal*, **56**, 979–980.

Gangopadhyay, P., Pyne, A.K., Maitra, D.N and Majumdar, S. (1981) Studies on biochemical constituents of blood with incorporation of neem seed cake (*Azadirachta indica*) in the ration of Murrah buffaloes. *Indian Journal of Animal Health*, **2**, 61–63.

Gupta, R.S. and Bhaid, M.V. (1981) Studies on agro-industrial by-products (deoiled neem fruit cake) in sheep feed consumption. *Indian Veterinary Journal*, **58**, 311–315.

Hentgen, A. (1985) Forage trees in India, a chance for rearing ruminants. *Fourages*, **101**, 105–119.

Ibrahim, I.A., Khalid, S.A., Omer, S.A. and Adam, S.E.I. (1992) On the toxicology of *Azadirachta indica* leaves. *Journal of Enthopharmacology*, **35**, 267–273.

Ibrahim, I.A., Khalid, S.A., Omer, S.A. and Adam, S.E.I. (1992a) On the toxicology of Azadirachta indica toxicosis in chicks. *Veterinary and Human Toxicology*, **34**, 221–224.

Jayal, M.M. (1963) Neem leaves as feed for livestock, their palatability, chemical composition and nutritive value. *Indian Veterinary Journal*, **40**, 283–289.

Katiyar, R.C., Sastry, V.R.B. and Agrawal, D.K. (1993) Nutrient utilization from alkali detoxified neem (*Azadirachta indica*) seed kernel cake, by cattle and buffalo. *Indian Journal of Animal Nutrition*, **10**, 223–226.

Ketkar, C.M. (1976) *Final Technical Report-Utilization of Neem (Azadirachta indica A. Juss) and its by products*. N.D. Sadhna Press, Pune, India.

Kumar, K.P., Reddy, C.R., Roa, M.R., Reddy, T.J., Rao, V.P. and Sastry, V.R.B. (1990) Nutrient digestibility in crossbred milch cows fed ration containing washed neem seed cake. *Indian Veterinary Journal*, **67**, 836–840.

Kumar, K.P., Reddy, C.R., Rao, M.R., Reddy, T.J., Rao, V.P. and Sastry, V.R.B. (1992) Utilization of water washed neem seed cake as cattle feed. *Indian Veterinary Journal*, **69**, 127–132.

Lehri, A., Gupta, A.R., Pathak, J.P. and Vasishtha, A.K. (1987) Quality oil and meal from need seed (*Azadirachta indica*). *Journal the Oil Technologist Association of India*, **19**, 94–98.

Mahendra, M., Reddy, C.R., Reddy, T.J., Rao, M.R. and Satyanaryana, A. (1995) Feeding dairy buffaloes with water washed neem seed cake—an economic ration. *Indian Veterinary Journal*, **72**, 543–545.

Maitra, D.N. and Duttagupta, R. (1982). Heat treatment of vegetable protein and its utilization by lactating buffaloes. *Indian Journal of Dairy Science*, **35**, 372–374.

Maitra, D.N., Roy, S. and Duttagupta, R. (1982) Efficiency of utilization of digestible and metabolised energy for milk production with neem cake. *Indian Journal of Dairy Science*, **35**, 368–372.

Murugan, M. and Kathaperumal, V. (1987) Utilization of tree fodder in sheep. *Indian Journal of Animal Sciences*, **57**, 1145–1147.

Murugan, M., Ravi, R. and Kathaperumal, V. (1987) Macro and micro mineral content in certain tree leaves of Tamilnadu. *Indian Journal of Animal Nutrition*, **4**, 126–128.

Nath, K., Agrawal, D.K., Hazan, Q.Z., Daniel, S.J. and Sastry, V.R.B. (1989) Water washed neem (*Azadirachta indica*) seed kernel cake in the feeding of milch cows. *Animal Production*, **48**, 497–502.

Nath, K., Rajagopal, S. and Garg, A.K. (1983) Water washed neem (*Azadirachta indica* A.Juss) seed kernel cake as cattle feed. *Journal Agricultural Science, UK*, **101**, 323–326.

Patel, B.M. and Patel, P.S. (1957) Fodder value of tree and vegetable leaves in Kaira District. *Indian Journal of Agricultural Sciences*, **27**, 307–315.

Patel, B.M. and Shukla, P.C. (1962) Nutritive value of neem leaves marygold and comfrey. *Indian Journal of Dairy Science*, **15**, 139–145.

Prakash, A.O., Mishra, A. and Mathur, R. (1991) Studies on the reproductive toxicity due to the extract of *Azadirachta indica* (seed) in adult cycle female rats. *Indian Drugs*, **28**, 163–169.

Punj, M.L. (1988) Availability and utilization of non-conventional feed resources and their utilisation by ruminants in South Asia residues: strategies for expanded utilization. Devendra, C. (ed.) *Proceedings of a Consultation held in Hissar*, India, 21–29, March, 1988.

Pyne, A.K., Mitra, D.N. and Gangopadhyay, P. (1979) Studies on the composition of milk with use of neem seed expeller cake (*Azadirachta indica*); on lactating buffaloes. *Indian Veterinary Journal*, **56**, 223–227.

Ramu, A., Reddy, T.J. and Raghavan, G.V. (1994) Water washed neem seed cake as substitute of deoiled rice bran in sheep and goat rations. *Indian Journal of Animal Nutrition*, **11**, 47–49.

Ramu, A., Reddy, T.J., Raghavan, G.V. and Djajanegara (1994a) Effect of feeding complete feeds containing water washed seed cake on nutrient utilisation in sheep and goat. Sukhamwati, A.(ed.) *Sustainable animal production and environment. Proceedings of the 7th AAAP Animal Science Congress*, Bali, Indonesia, 11–16 July, 1994.

Ranjhan, S.K. (1980) *Animal Nutrition and Feeding Practice in India* Vikas Publications, New Delhi.

Rao, P.V. (1987) Chemical composition and biological evaluation of debittered and defatted neem (*Azadirachta indica*) seed kernel cake. *Journal American Chemical Society*, **64**, 1348–57.

Reddy, V.R. and Rao, P.V. (1988) Utilization of undecorticated expeller-processed or solvent extracted neem cake in chicks. *Indian Journal of Animal Sciences*, **58**, 835–839.

Reddy, V.R. and Rao, P.V. (1988a) Utilization of differently processed undecorated neem cake in broiler chicks. *Indian Journal of Animal Sciences*, **58**, 840–842.

Reddy, V.R. and Rao, P.V. (1988b) Utilization of chemically treated neem cake in broilers. *Indian Journal of Animal Sciences*, **58**, 958–963.

Reddy, V.R., Rao, P.V. and Reddy, C.V. (1988) Utilization of chemically treated neem oil in broiler chicks. *Indian Journal of Animal Sciences*, **58**, 830–834.

Reddy, V.R., Rao, P.V. and Reddy, C.V. (1988a) Chemical composition and nutritive value of processed neem cake. *Indian Journal of Animal Sciences*, **68**, 870–873.

Reddy, T.J., Ramu, A. and Raghavan, G.V. (1994) Nutrient utilisation by goat fed complete feeds incorporated with water washed neem seed cake. *Indian Journal of Animal Production and Management*, **10**, 81–84.

Rukmani, C. (1987) Chemical and nutritional evaluation of neem oil. *Food Chemistry*, **26**, 119–124.

Rukmani, C., Rao, P.U. and Raychaudhuri, S.P. (1991) Chemical composition and biological evaluation of debitterised neem oil and neem cake. *Recent Advances in Medicinal, Aromatic, and Spice Crops, Vol. I. International Conference*, New Delhi, 28–31 January, 1989, pp. 191–196.

Salawu, M.B., Adedeji, S.K. and Hassan, W.H. (1994) Performance of broilers and rabbits given diet containing full fat neem (*Azadirachta indica*) seed meal. *Animal Production*, **58**, 285–289.

Sastry, V.R.B. and Agrawal, D.K. (1992) Utilization of water washed neem (*Azadirachta indica*) seed kernel cake as a potential source for growing pigs. *Journal of Applied Animal Research*, **1**, 103–107.

Singh, N.P. and Pathak, B.C. (1981) A note on the palatability of neem tree leaves for sheep. *Food, Farming and Agriculture*, **14**, 47–48.

Singh, R.K. (1982) *Fodder Trees in India*. Oxford and IBH Publishing Co., New Delhi, India.

Singh, Y.P., Bahga, H.S. and Vijjan, V.K. (1985) Toxicity of water extract of neem (*Azadirachta indica* A. Juss) in poultry birds. *Neem Newsletter*, **2**, 17–18.

Singhal, K.K. and Mudgal, V.D. (1983) Versatile neem (Azadirachta indica) a review. *Agriculture Review*, **4**, 1–10.

Shukla, P.C. and Desai, M.C. (1988) Neem (*Azadirachta indica*) as a source of cattle feed. *International Tree Crops Journal*, **5**, 135–142.

Thomas, C.P and Prasad, D.A. (1983) Effect of economic ration on the performance, nutrient utilization and carcass quality of white Yorkshire Pigs. *Indian Journal of Animal Sciences*, **53**, 738–742.

Vasishtha, A.K., Pathak, J.P. and Lehri, A. (1992) Studies on solvent extraction of neem seed (*Azadirachta indica*). *Journal of the Oil Technological Association of India*, **24**, 75–80.

Verma, A.K. (1995) Feeding of water washed neem (*Azadirachta indica*) seed kernel cake to growing goats. *Small Ruminant Research*, **15**, 105–111.

Verma, A.K., Sastry, V.R.B. and Agrawal, D.K. (1996) Chevan characteristics of goats fed diets with water washed neem (*Azadirachta indica*) seed kernel cake. *Small Ruminant Research*, **19**, 55–61.

Vijjan V.K. (1983) Note on the removal of growth depressant factor in neem (*Azadirachta indica*) seed cake. *Journal of Veterinary Physiology and Allied Sciences*, **2**, 25–28.

Vijjan, V.K. and Parihar, N.S. (1983) Toxic effects of neem (*Azadirachta indica*) seed cake feeding to rats. *Journal of Environmental Biology*, **4**, 39–41.

Vijjan, V.K., Tirpathi, H.C. and Parihar, N.S. (1982) Note on the toxicity of neem (*Azadirachta indica*) seed cake in sheep. *Journal of Environmental Biology*, **3**, 47–52.

Webb, R. (1988) A preliminary investigation into the fodder qualities of some trees in Sudan. *International Tree Crops Journal*, **5**, 1–2, 9–17.

Zech, W. and Weinstable, P.E. (1983) Location, state of nutrition and feed value of tree species important in forestry in Upper Volta. *Plants Research and Development*, **17**, 42–60.

17. NEEM AND POLLUTION

Rapid industrialization, urbanization, and congestion of population in a few pockets, in most part of the world, are giving rise to pollution caused by emission of gases such as carbon monoxide, carbon dioxide, sulfur dioxide and nitrogen peroxide which may play havoc with the human population. In Indian culture, neem has been referred as an "air purifier" so it may be an avenue tree of choice in thickly populated areas, by its capacity to survive in adverse conditions, absorb some of the environmental pollutants, and act as an "air freshener" by releasing oxygen and mild odorous principles.

INDUSTRIAL POLLUTION

Tanneries

Tannery is one of the industries responsible for pollution of river water. In third world countries, in some areas, the cattle population exceeds that of humans, so an appreciable amount of animal hide is available which is treated with tanning materials to turn it into leather. The whole process requires repeatedly washing with water, so the water requirement is very high; after washing, this water becomes heavily contaminated and is drained back to the rivers.

Chaturvedi (1986) tested neem as one of the trees for tolerance to tannery waste water. The survival rate of the tree was 22–94 percent. Ramanuja and Misra (1986) did culture experiments over 12 weeks, with 8–9 month old seedlings. Plants were irrigated weekly with 2 liters of effluent from a tannery settling tank. Neem was found quite tolerant to this waste water.

The other major polluting industries are thermal plants and chemical factories, such as those for fertilizers and pesticides, which release carbon dioxide, sulfur dioxide and nitrogen peroxide, in addition to suspended particles like dust or fly ash.

Chemical Factories

Devi and Patel (1982) and Patel and Devi (1985) studied morphological variation in the vegetation around a fertilizer complex. The variation noted was reduced foliage or defoliation, cracking and peeling of bark, reduced leaf and leaflet area and petiole length, mutilation of leaves in various ways and the total absence of flowering.

In the case of leaves in normal plants, the cell wall was undulating but in the polluted area it was straight, with reduced stomatal frequency and some variation in stomatal width and pore area. Starch, insoluble polysaccharides and lipids also varied.

On 3 December 1984, large quantities of methyl iso-cyanate (MIC) escaped from a pesticide plant in central India. Ram Prasad and Pandey (1985) studied the effect of this poisonous gas on the neem tree. The tree was sensitive to MIC. There was defoliation and blackening of the foliage, but the tree revived two months after the injury with the emergence of new leaves. Farooq et al. (1988) studied the sulfur dioxide resistance of trees and also visible symptoms of sulfur dioxide absorption. Seedlings of 12 species, including neem, were exposed to sulfur dioxide to various concentrations and it was found that neem could tolerate this gas to a major extent. Rao and Dubey (1990), in their study to find differential responses to sulfur dioxide, analyzed the neem tree along with others for stomatal conductance, sulphate, protein, superoxide dismutase and peroxides for one year in an ambient environment with varying concentration of sulfur dioxide. The results indicated that trees under sulfur dioxide stress developed phytotoxicity by undergoing certain biochemical changes.

To find out the effect of sulfur dioxide exposure to tree saplings, Krishnnayya and Bedi (1989) noted that this gas damaged chloroplasts and cytoplasm in palisade cells, followed by rupturing of the outer envelope of the chloroplasts and extrusion of plastoglobuli and starch into the cytoplasm. Thinma Raju *et al.* (1993) observed that a neem tree growing in a highly polluted area was not affected by various gases, whereas some other trees exhibited symptoms of defoliation, die-back, poor flowering and fruiting.

Thermal Stations

In one experiment, fly ash-induced injury to leaves and proline metabolism in plants growing at two different distances away from a thermal powerhouse was studied. Trees closer to the source of pollution had higher dust deposition, leaf injury and proline accumulation at a lower pH of cell sap as compared to neem trees away from the source. Proline accumulation was present at both sites, throughout the period of study of four months. It indicated the greater ability of neem to adapt to stress from exposure to air pollution (Chauhan and Varshney, 1989; Maini and Harapanahalli 1991; Anonymous, 1991).

Beg *et al.* (1990) studied the performance of neem trees at five selected sites. In the vicinity of a power station, gases like nitrogen peroxide and sulfur dioxide were below the permissible levels, but the total suspended particles were higher. Air pollution had a major effect on chlorophyll. It was seen in this study that there was some destruction of chlorophyll, showing thereby that neem is moderately sensitive when exposed to sulfur dioxide in various concentrations but could tolerate this gas to a major extent.

Exhaust from Automobiles

Pollution due to the exhaust from automobiles in some congested areas is fairly high. The effect of this exhaust on the survival of neem, if planted as an avenue tree, is very important. Keeping this point in view, Bhatti and Iqbal (1988) studied the leaf length area, dry weight, etc. of these trees. It was noted that neem tolerated this type of pollution very well and can be planted as a roadside tree in thickly populated areas.

When dust loading of leaves in an automobile polluted area was studied, it was seen by Satyanarayana *et al.* (1990) that the neem tree accumulated a thick, sticky crust. In the leaves, as compared to the control, the epidermal cell size was smaller and stomatal frequency was higher. Sharma and Roy (1995) also observed the same features on leaves when subjected to automobile exhaust.

Steel Industry

The phytotoxic effects of aerial discharge from the steel industry were evaluated for neem by Kumawat and Dubey (1988). In the neem trees growing in the area, the chlorophyll pigments, carotenoids and leaf pH level decreased, while the leaf injury index, leaf area/dry weight ratio, conductivity of leaf disk water, sulfate content and total chlorophyll:sulfate ratio increased. The authors observed that the pollution injury was maximum in winter, followed by summer and the rainy season.

Detection of Heavy Metals

The neem bark was tested to monitor heavy metals in polluted sites, and compared with those absorbed by moss. Lead, zinc and iron content in both materials were higher at most sites (Kakulu, 1993).

Water Purification

Johri *et al.* (1993) developed a process of flocculation of water pollutants by using extract of seeds of neem along with *Moringa oleifera* and *Madhuca latifolia*. This extract formed a floc with the contaminants and the suspended particles settled down, giving rise to clean water.

ENVIRONMENTALLY FRIENDLY

Bees

Before neem products could be used against insects and pests, quite exhaustive studies were carried out on the safety aspects, particularly with regard to warm-blooded animals. Concern was also expressed over its effect on honey bees. Schmutterer and Holst (1987) studied this aspect. The neem-treated flowers were not repulsive to bees and the neem preparations did not cause any serious damage to their systems. On the other hand, neem controlled the tracheal mites of the bee (Liu, 1995).

Neem-based products were not only found safe for earthworms, but for young salmon also (Wan *et al.*, 1996).

TOXICITY

Neem trees in a grove are eco-friendly, but pollens have been known to cause allergy in some cases. Karmakar and Chatterjee (1994) isolated and characterized IgE-reactive proteins for pollens. They have shown that AIaI and AIaIVb are the major allergens. Amino acid analysis of these, the effect of pH on them and cross-activity has also been carried out by these authors.

REFERENCES

Anonymous (1991) Fly ash induced foliar injury and proline metabolism in plant growing near power house. *Advance in Plant Science*, **4**, 298–303.

Beg, M.U., Far, O.M., Bhargava, S.K., Kidwai, M.M. and Lal, M.M. (1990) Performance of trees around a thermal power station. *Environment and Ecology*, **8**, 791–797.

Bhatti, G.H. and Iqbal, M.Z. (1988) Investigations into the effect of automobile exhausts on the phenology, periodicity and productivity of some roadside trees. vide *Forestry Abstract* (1990), 051–02158.

Chaturvedi, A.N. (1986) Trees and shrubs for control of tannery waste water in India. *Environmental Conservation*, **13**, 164–165.

Chauhan, A. and Varshney, C.K. (1989) Alteration in the buffering capacity of leaves of tree species growing near a coal fired thermal power station, New Delhi. *International Journal of Ecology and Environmental Sciences*, **15**, 117–124.

Devi, G.S. and Patel, J.D. (1982) Variation in the vegetation around a fertilizer complex. I. Morphological Variation. *Indian Botanical Reporter*, **1**, 114–118.

Farooq, M., Saxena, R.P. and Beg, M.U. (1988) Sulfur dioxide resistance of Indian trees. I. Experimental evaluation of visible symptoms and SO2 sorption. *Water, Air, Soil Pollution*, **40**, 307–316.

Johri, P.K., Johri, R., Lal, N. and Katiyar, A.K. (1993) Flocculation of water pollutants through plant origin materials. *Proc. World Neem Conference*, Bangalore, India, 24–28 Feb., 1993.

Kakulu, S.E. (1993) Biological monitoring of atmospheric trace metal deposition in North Eastern Nigeria. *Environmental Monitoring and Assessment*, **26**, 137–43.

Karamakar, P.R. and Chatterjee, B.P. (1994) Isolation and characterisation of two IgE-reactive proteins from Azadirachta indica. *Molecular and Cellular Biochemistry*, **131**, 87–96.

Krishnnayya, N.S.R. and Bedi, S.J. (1980) Effect of sulfur dioxide and ascorbic acid on the plastid ultra structure of Azadirachta indica, leaves. *Annals of Botany*, **64**, 311–313.

Kumawat, D.M. and Dubey, P.S. (1988) Steel industry aerial discharges and response of two tree species. *Geobios, Jodhpur*, **15**, 176–180.

Liu, T.P. (1995) The controlling tracheal mites in colonies of honey bees with neem (Margosan-0) and flumethrin (Bayvarol). *American Bee Journal*, **135**, 562–566.

Maini, A. and Harapanahalli, A.B. (1991) Environmental impact due to Satpura thermal power station. *Journal of Economic Botany and Phytochemistry*, **2**, 33–34.

Patel, J.D. and Devi, G.S. (1985) Studies on leaf epidermis of some angiosperm species growing under pollution stress of fertilizer complex. *Journal of Plant Anatomy and Morphology*, **2**, 1–10.

Ram Prasad and Pandey, R.K. (1985) Methyl isocynate (MIC) hazard to the vegetation of Bhopal. *Journal Tropical Forestry*, **1**, 40–50.

Ramanuja, S. and Misra, G.M. (1986) Impact of tannery effluents on different forest species. *Van Vigyan*, **24**, 37.

Rao, M.V. and Dubey, P.S. (1990) Explanation for the differential response of certain tropical trees species to SO under field conditions. *Water, Air, Soil Pollution*, **51**, 297–305.

Satyanarayana, G., Pushpalatha, K. and Acharya, U.H. (1990) Dust loading and leaf morphological trait changes of plants growing in automobile polluted area. *Advances in Plant Sciences*, **3**, 125–130.

Schmutterer, H. and Holst, H. (1987) On the effect of enriched and formulated neem seed kernel extract AZT-VR-K on the honey bee (in German) *Zeitschrift fur Angewandte Entomologie*, **103**, 208–213.

Sharma, M. and Roy, A.N. (1995) Effect of automobile exhaust on the leaf epidermal features of Azadirachta indica and Dalbergia sisoo. *International Journal of Mendelism*, **12**, 18–19.

Thinma Raju, K.R., Reddy, T.V. and Chadndra Gowda, M. (1993) Indian Neem (Melia azadirachta) an avenue tree. *Proc. World Neem Conference*, Bangalore, India, 24–28 Feb., 1993.

Wan, M.T., Watts, R.G., Isman, M.B. and Strub, R. (1996) Evaluation of the acute toxicity to juvenile Pacific Northwest salmon of azadirachtin, neem extract and neem based products. *Bulletin of Environmental Contamination and Toxicology*, **56**, 432–439.

18. NEEM AND HOUSEHOLD PESTS

Various studies have now indicated that neem may be useful against household and kitchen garden pests, as follows.

ANT

The sterilizing effect of neem extract on the queen and workers of *Formica polyctena* were tested by feeding and fumigation. Fumigation increased egg production, but in feeding experiments the laying capacity of eggs was reduced with higher concentration of extract (Schmidt and Pesel, 1987).

BED BUG

Toxicity against *Cimex lectularius* has been found (Naqvi *et al.*, 1993).

BIRDS

Enormous loss to grains and young seedlings is caused by birds. Syamsunder Rao *et al.* (1993) tried various commercial neem formulations to find a non-lethal, environmentally safe product. The common birds showed feeding aversion and avoided consumption of neem-treated grains.

COCKROACH

A commercial preparation from neem seed extract was tried by Adler and Uebel (1985) against six species of cockroach, for toxicant, growth inhibitor and repellent action. Last instar nymphs showed increased mortality and retarded development but the other actions differed in different species. Topical application or an injection was toxic but the surface treated with neem extract was not effective in controlling them.

GOAT

The stray or wild herbivorous animal causes enormous loss to kitchen gardens and avenue trees, and protection of these can only be provided by fencing which is not only expensive but very often not effective. If some deterrent is applied on these, which is not toxic to vegetation and safe for the environment, most of these plants can be saved from these grazing animals. Gope *et al.* (1988) sprayed neem preparations on tea bushes using hand sprayers; goats avoided the sprayed plants and after three months there was considerably less damage to the bushes.

This treatment may also be useful against deer, which in the USA not only cause extensive damage to plants but also spread Lyme disease.

HOUSE FLY

Ethanolic extract of neem was used by Azmi *et al.* (1995) against *Musca domestica*. It increased mortality and was comparable to deltmethrin. Azadirachtin inhibited moulting in the larvae of the face fly *M. autumnalis* (Gaaboub and Hayes, 1984). A compound from neem extract caused morphogenic effects on various stages of fly, including weight reduction and abnormal development (Naqvi *et al.*, 1995).

MOSQUITO

For control of malaria, Japanese encephalitis and dengue fevers, eradication of the mosquito is very important. The mosquito has spread to new areas because of the construction of dams, canals, and other irrigation facilities. The improper disposal of drainage water has provided further breeding grounds. The recently synthesized insecticides were indiscriminately used for the control of the mosquito, but some strains developed resistance against these chemicals. In due course of time, the polluting effects of these chemicals became well known, and a search started for some eco-friendly agent such as neem.

Neem is effective against the mosquito in two ways, as a larvicide, and as a repellent.

Larvicide

During and after the second world war, kerosene oil and tar were commonly used for larvicidal effect. It was a common practice to sprinkle these non-biodegradable products on standing water or in other mosquito breeding grounds in mosquito-infested areas. Recently, it has been observed that eco-friendly neem extract and neem oil can replace the above petroleum products (Attri and Ravi Prasad, 1980). Petroleum ether extract of neem was tested by Deshmukh and Renapurkar (1987) against third instar larvae of *Culex quinquefasciatus*. It totally suppressed larval development. The essential oil obtained by steam distillation had good activity against larvae of *Anopheles stephensi* (Kumar and Dutta, 1987).

Four azadirchtin-rich fractions also had this effect (Rao *et al.*, 1988). Petroleum ether extract of dried leaves (total alkanes) was tried for larvicidal action. A one percent solution of extract gave 100 percent mortality. A mixture of purified extract was more effective than the crude leaf extract (Chavan and Nikam, 1988).

The larvicidal effect of neem seed bitters was also tested by Rao *et al.* (1989), who observed that 2000 ppm of the bitter produced 100 percent mortality in 72 hours. Reuben *et al.* (1990) discussed at length the use of neem extract for biological control of the mosquito. Tare and Sharma (1991) and Rao *et al.* (1992) experimented with neem seed bitter principles at different concentrations against newly moulted fourth instar larvae of *C. quinquefasciatus* in flooded rice fields. The bitter principles were very effective, even at a concentration of 2000 ppm. There was 100 percent mortality in 72 hours. The neem extract was applied to the rice field for the dual purposes of controlling the mosquito vector of Japanese encephalitis virus and enhancing the grain yield. The standardized extract of neem was found effective for both purposes.

In another study on larvicidal potential, *Aedes aegypti* was found to be more susceptible than *C. quinquefasciatus* (Monzon *et al.*, 1994) to neem. When larvae of *A. aegypti* were reared in water containing different concentrations of a commercial preparation having 40 percent azadirachtin, it was seen by Boschitz and Grunewald (1994) that sensitivity to the product decreased with increasing age of the larvae. Female mosquitoes laid fewer eggs which were directly proportional to increasing concentration of the neem preparation to which they had been exposed during larval development.

The crude extracts of neem were effective against mosquito larvae but these extracts had a problem of stability and storability, when used on a large scale in the rice fields. Rao *et al.* (1995) tried a good quality neem product for control of larval cluicine mosquitoes. It also produced a slight but significant reduction in the population of anopheline pupae.

Mittal *et al.* (1995) tested six neem products against fourth instar larvae of *A. stephensi*, *C. quinquefasciatus* and *A. aegypti*. Larvae of *A. stephensi* were the most susceptible, those of *A. aegypti* the least. Jin Ping *et al.* (1995) studied toxicity and the growth-regulating activity of neem seed kernel extract in the larvae of *C. quinquefasciatus*. This extract induced prolongation of first instar larvae but caused death and morphogenetic aberrations of fourth instar larvae.

Neem oil and deoiled cake were also found to be promising larvicides by Amorose (1995). Third instar larvae were more susceptible than fourth instar larvae. Nagpal *et al.* (1995) soaked wooden balls in 5, 10, 20 percent neem oil diluted in acetone for the control of *Anopheles stephensi* and *Aedes egypti*. These balls were dipped in storage and overhead tanks. The balls soaked in 5 percent neem oil gave the best results. Similarly, pieces of cardboard dipped in oil may be used as a mosquito-repellent mat.

Repellent

Lemon grass oil in various forms has been used as a mosquito repellent. It has been used as an active ingredient of ointments or oils for external application on the exposed part of the body or in fumigants, coils, mats and candles, for fumigation. Lemon grass is a volatile oil, so preparations containing it are to be applied repeatedly to the body or the surroundings should be continuously fumigated. Recently, in mats, lemon grass oil has been replaced by synthetic pyrethroids and other compounds, because of their persistence nature, but some studies indicate that these are hazardous, particularly for infants and children in an enclosed atmosphere.

In India, for repelling insects or for so-called purification of air, premises are often fumigated with a mixture of neem leaves, oleo-gum resin, particularly that of *Commiphora wighti*, and sulfur. Pandian *et al.* (1989) and Pandian & Manoharan (1995) observed that with smoke from dry leaves of neem, the landing and biting rates of mosquitoes were reduced considerably.

Sharma et al. (1993, 1995) mixed 2 percent of neem oil in coconut oil and applied this mixture of oils to the exposed part of human volunteers. This provided complete protection against mosquito bites for 12 hours. In a field trial in a village in west India, a mixture of 0.5, 1 or 2 percent neem oil in coconut gave 79.65, 96.07, and 98.03 percent protection respectively against *Anopheles culicifacies* in an all-night biting test. Two percent neem oil provided 75 percent protection against other types of mosquito (Kant and Bhatt, 1994). Similar results were obtained by Mishra et al. (1995) with 1–4 percent neem oil in coconut oil, when applied to human volunteers in a tribal village. Dua et al. (1995) applied a neem cream to see if it can provide protection against mosquitoes. One application of the cream was effective in 68 percent of the population for four hours.

SAND FLY

Two percent neem oil mixed in coconut or mustard oil provided 100 percent protection against *Phlebotomus argentipes* throughout the night under field conditions. It was effective against *P. paptsi* for about seven hours only (Sharma and Dhiman, 1993).

SNAIL

The snails *Lymnaea acuminata* and *Indoplanorbis exustsus* are hosts of *Fasciola gigantica*, which causes fascioliasis. *Melania scabra* is also a vector snail (Muley, 1978). The various types of neem products alone or in a mixture (Bali et al., 1985) or with *Cedrus deodara* and *Embelia ribes* have shown molluscicidal activity (Singh et al., 1995) and may be effective against the diseases caused by the snails. The toxic effects of pure azadirachtin against snails are greater as compared to other neem products (Singh et al., 1996).

TERMITE

Neem cake seems to act as a repellent barrier against *Odontotermes* spp. and *Microtermes obesi* (Gold et al., 1989). A methanolic extract of neem oil with antifeedant activity was tried against the termite *Reticulitermes speratus* and was found effective (Ishida et al., 1992).

TICK

Azadirachtin inhibited the onset of oviposition by only a few days in the case of the tick, *Amblyomma americanum* (Lindsay and Kaufman, 1988).

REFERENCES

Adler, V.E. and Uebel, E.C. (1985) Effect of formulation of neem extract on six species of cockroaches (Orthoptera: Blaberidae, Blattidae and Blattellidae). *Phytoparasitica*, **13**, 3–8.

Amorose, T. (1995) Larvicidal efficacy of neem (Azadirachta indica L) oil and defatted cake on Culex quinquefasciatus. *Geobios*, 22, 169–173.

Attri, B.S. and Ravi Prasad (1980) Neem oil extractive—an efficient mosquito larvicide. *Indian Journal of Entomology*, 42, 371–374.

Azmi, M.A., Naqvi, S.N.H., Khan, M.F., Naz, S. and Akhtar, K. (1995) Toxicity of neem seed extract (NSPE) against Musca domestica as compared to deltmethrin. *Geobios*, 22, 229–230.

Bali, H.S., Sawai Singh and Pati, S.C. (1985) Preliminary screening of some plants for molluscicidal activity against two snail species. *Indian Journal of Animal Science*, 55, 338–340.

Boschitz, C. and Grunewald, J. (1994) The effect of Neem Azal on Aedes aegypti (Diptera: Cullicidae). *Applied Parasitology*, 35, 251–256.

Chavan, S.R. and Nikam, S.T. (1988) Investigation of alkanes from neem leaves and their mosquito larvicidal activity. *Pesticide, Bombay*, 22, 32–33.

Deshmukh, P.B. and Renapurkar, D.M. (1987) Insect growth regulator activity of some indigenous plant extracts. *Insect Science and its Application*, 8, 81–83.

Dua, V.K., Nagpal, B.N. and Sharma, V.P. (1995) Repellent activity of neem cream against mosquitoes. *Indian Journal of Malariology*, 32, 47–53.

Gaaboub, I.A. and Hayes, D.K. (1984) Biological activity of azadirachtin, component of the neem tree inhibiting molting in the face fly, Musca autumnalis De Geer (Diptera: Muscidae). *Environmental Entomology*, 13, 803–812.

Gold, C.S., Wightman, J.A. and Pimpert, M. (1989) Mulching effects on termite scarification of drying groundnut pods. *International Arachis Newsletter*, 6, 22–23.

Gope, B., Mukherjee, S. and Das, S.C. (1988) Neem oil cake as goat repellent. *Two and a Bud*, 35, 48–49.

Ishida, M., Serit, M., Nakata, K., Juneja, L.R., Kim., M. and Takahashi, S. (1992) Several antifeedants from neem oil, Azadirachta indica A. Juss, against Reticulitermes speratus Kolbe (Isoptera: Rhinotermitidae). *Bioscience, Biotechnology, and Biochemistry*, 11, 1835–1838.

Jin-Ping, Pan-Xiaojun and Zhao-Shan Huan (1995) Toxicity and growth regulating activity of neem seed kernel extract (AZAL-S) to the larvae of Culex quinquefasciatus. *Entomologia Sinica*, 2, 64–69.

Kant, R. and Bhatt, R.M. (1994) Field evaluation of mosquito repellent action of neem oil. *Indian Journal of Malariology*, 31, 122–125.

Kumar, A. and Dutta, G.P. (1987) Indigenous plant oils as larvicidal agent against Anopheles stephensi mosquitoes. *Current Science, India*, 56, 959–960.

Lindsay, P.J. and Kaufman, W.R. (1988) The efficacy of azadirachtin on putative ecdysteroid sensitive systems in the ixodid tick Amblyomma americanum L. *Journal of Insect Physiology*, 34, 439–442.

Mishra, A.K., Singh, N. and Sharma, V.P. (1995) Use of neem oil as a mosquito repellent in tribal villages of Mandla District, Madhya Pradesh, India. *Indian Journal of Malariology*, 32, 99–103.

Mittal, P.K., Adak, K. and Sharma, V.P. (1955) Bioefficacy of six neem (Azadirachta indica) products against mosquito larvae. *Pesticide Research Journal*, 7, 35–38.

Monzon, R.B., Alvior, J.P., Luczon, L.L.C., Morales, A.S. and Mutuc, F.E.S. (1994) Larvicidal potential of five Philippine plants against Aedes aegypti (Linnaeus) and Culex quinquefasciatus *Southeast Asian Journal of Tropical Medicine and Public Health*, 25, 755–759.

Muley, E.V. (1978) Biological and chemical control of the vector snail Melania scabra (Gastropoda: Prosobranchia). *Bulletin Zoological Survey (India)*, 1, 1–5.

Nagpal, B.N., Srivastava, A. and Sharma, P.V. (1995) Control of mosquito breeding using wood scrapings treated with neem oil. *Indian Journal of Malariology*, 32, 64–69.

Naqvi, S.N.H., Jahan, M., Tabassum, R., Qamar, S.J. and Ahmed, I. (1995) Toxicity and teratogeny caused by Coopex 25EC and a neem extract (N-7) against 3rd instar larvae of Musca domestica L. *Pakistan Journal of Zoology*, 27, 27–31.

Naqvi, S.N.H., Tabassum, R., Nurulain, S.M. and Khan, M.F. (1993) Comparative toxicity of RB— (a neem formulation) and malathion against bed bugs (Cimex lectularius) in laboratory and field conditions. *Proc. of Pakistan Congress of Zoology*, **13**, 369–377.

Pandian, R.S., Dwarkanath, S.K. and Martin, P. (1989) Repellent activity of herbal smoke on the biting activity of mosquitoes. *Journal of Ecobiology*, **1**, 87–89.

Pandian, R.S. and Manoharan, A.C. (1995) Herbal smoke: a potential repellent and adulticide for mosquitoes. *Insect Environment*, **1**, 14–15.

Rao, D.R., Reuben, R., Gitanjali, Y., Srimannarayana, G. and Raghunatha-Rao, D. (1988) Evaluation of four azadirachtin rich fractions from neem Azadirachta indica A. Juss (family: Meliaceae) as mosquito larvicide. *Indian Journal of Malariology*, **25**, 67–72.

Rao, D.R., Reuben, R. and Nagasampagi, B.A. (1995) Development of combined use of neem (Azadirachta indica) and water management for the control of culcine mosquitoes in the rice fields. *Medical and Veterinary Entomology*, **9**, 25–33.

Rao, D.R., Reuben, R. and Saxena, R.C. (1989) Larvicidal activity of neem seed bitters (NSB) against Culex quinquefasciatus in flooded rice fields. *International Rice Research Newsletter*, **14**, 28.

Rao, D.R., Venugopal, M.S., Nagasampagi, B.A. and Schmutterer, H. (1992) Evaluation of neem, Azadirachta indica, with and without water management, for the control of culcine mosquito larvae in rice fields. *Medical and Veterinary Entomology*, **6**, 318–324.

Reuben, R., Raghunatha, Rao, D., Sebastian, A., Corbet, P.S., Wu, N., Liao, G.H. and Rao, D.R. (1990) Biological control methods for community use. In C.F. Curtis (ed.) *Appropriate Technology in Vector Control*. CRC Press, Inc., Florida, USA, pp. 139–158.

Schmidt, H.H. and Pesel, E. (1987) Studies of the sterilising effect of neem extract in ants. Proceedings of the 3rd International Neem conference, Nairobi, Kenya, 10–15 July, 1986. Schmutterer, H., Ascher, K.R.S. (eds.) Eschborn, D.G.T.Z. (Germany).

Sharma, S.K., Dua, V.K. and Sharma, V.P. (1995) Field studies on the mosquito repellent action of neem oil. *Southeast Asian Journal of Tropical Medicine and Public Health*, **26**, 180–181.

Sharma, V.P., Ansari, M.A. and Razdan, R.K. (1993) Mosquito repellent action of neem (Azadirachta indica) oil. *Journal of the American Mosquito Control Association*, **9**, 359–360.

Sharma, V.P. and Dhiman, R.C. (1993) Neem oil as a sand fly (Diptera: Psychodidae) repellent. *Journal of the American Mosquito Control Association*, **9**, 364–366.

Singh, K., Singh, A. and Singh, D.K. (1995) Molluscicidal activity of different combinations of the plant products used in the molluscicide Pestoban. *Biological Agriculture and Horticulture*, **12**, 253–261.

Singh, K., Singh, A. and Singh, D.K. (1996) Molluscicidal activity of neem (Azadirachta indica A. Juss). *Journal of Ethnopharmacology*, **52**, 35–40.

Syamsunder Rao, P., Seshachalam Babu, K., Anand Rao, M. and Vardhani, B.P. (1993) Use of certain botanical formulations as potential repellents in the bird pest management. *Proc. World Neem Conference*, 24–28 February, 1993, Bangalore, India.

Tare, V. and Sharma, R.N. (1991) Larvicidal activity of some tree oils and their common chemical constituents against mosquitoes. *Pesticide Research Journal*, **3**, 169–172.

19. PROTECTION OF FOOD MATERIALS

TRADITIONAL USES

It was a common practice, particularly in rural areas, to put dry leaves of neem between folds of dry cloth, or in stored grains and cereals to ward off various insects. In some parts of south India and Sri Lanka, fumigation with neem leaves was practiced instead. Vijayalakshmi *et al.* (1995) have described simple methods of treating gunny bags, storage bins and rooms with a solution of neem kernel powder or plastering the wall of bamboo baskets with a paste of cow dung and neem cake powder. Wealth of India (1948) gave a brief account of the use of neem in the protection of food items in daily use, including potato, without cold storage. It has been seen by experience that various treatments of neem products on cereals and legumes are harmless and do not effect palatability or cooking quality.

According to Islam (1993), neem may be used in Bangladesh for the protection of food materials from insects and worms during storage, in the following ways in the rural areas:

1. Fresh or dried leaves mixed with grains.
2. Spraying of neem oil in the storage area.
3. Rubbing a paste of seed or oil on the inside walls of the basket.
4. Plastering the hut with a mixture of clay, cow dung and neem seed powder.
5. Soaking the bags in the crude extract of neem seed.
6. Hanging of leaves or making a leaf mat in the storage room.
7. Fumigation with a mixture of neem leaves and oleo resins.
8. Washing or rinsing fresh fish with a water extract of neem.
9. Hanging bunches of leaves over the food or meat to repel flies.

Neem protects the stored grains and fresh food in various ways but it is not effective against all the causative organisms. It may be acting differently on different insects. In some insects it may prevent oviposition to varying degrees, may halt post-embryonic development completely and the emergence of the insects may be delayed for months. In others it deterred feeding and the worm died of starvation. In general, seed extract is more effective than the leaves. It suppressed the growth of fungi producing aflatoxins.

EFFECT ON CEREALS

In Nigeria, Ivbijaro (1983) mixed maize grain with dry ground seed of neem. It saved the maize from damage by the weevils *Sitophilus oryzae* for six months. Adult weevils placed on the maize treated earlier with neem had a very high mortality rate. Zhang and Zhao (1983) in China found that neem oil (5 ml/kg), when mixed in stored rice, reduced the population growth of weevils considerably. In India, Jadhav and Jadhav (1984) observed that neem oil in various concentrations significantly inhibited the emergence of the pulse beetle *Callosobruchus maculatus* in chick pea (gram). Malik *et al.* (1984) studied the anti-feedant and repellent properties of neem. Pandey *et al.* (1986) found the same effect with petroleum ether extract of neem on *C. chinensis*. Singh and Kataria (1986) experimented with the effect of deoiled neem kernel powder and leaf powder on the development of *Trogoderma granarium*. All the larvae died but the ethanolic extract showed higher toxicity as compared to the kernel powder; the neem leaf was least effective.

Singh *et al.* (1987) observed the activity of extract of neem against the insect *Rhyzopertha dominica* in stored grains. Gupta *et al.* (1988) tried neem and some other non-edible oils on storage of wheat seed and their germinability. Oils of neem and *Butea frondosa* at the rate of 5 ml/kg were most effective. The germination of seed was not affected by this treatment. Das (1989) showed that 1 ml/100 gm neem was an effective surface protectant against *C. chinensis* in chick pea for six months. Cobbinah and Appiak-Kwartang (1989) compared various neem products, along with others, against *Sitophilus zeamais*. The damage was less to maize treated with neem oil or neem wood ash as compared to the untreated one. But ash was less effective. Kossou (1989) also tried various plant parts like neem seed kernel, leaf, flower and bark against *S. zeamais* in maize.

The activity of neem extract of water and methylated spirit (commercial ethanol, mixed with methanol and the other solvents to make it unfit for human consumption) on *C. maculatus, S. oryzae, S. zeamis* and *Cylas puncticollis* for the protection of cow pea (*Vigna unguiculata*) and maize was studied. The extracts showed more activity in cow pea as compared to maize, and suppressed *C. maculatus* more than *S. oryzae*. There was no effect on *C. puncticollis* (Makanjuola, 1989).

Jilani and Saxena (1990) tested the repellence of neem oil and neem based insecticide against *R. dominica* for eight weeks and found that these have a long range of effect because of persistence. Dabire (1992) tested the traditional claim about the effect of neem seed cake on the protection of cow pea during storage.

Kumar and Mehta (1993) found 4 percent neem leaves very effective against *R. dominica* in milled rice. High mortality of *R. dominica* and *Tribolium castaneum* by neem oil was also seen by Mohiuddin *et al.* (1993), while Kahare *et al.* (1993) tried 1 percent neem extract against *C. chinensis* and found considerable reduction in the number of eggs of this bruchid.

Dust of deoiled neem kernel was tested against three pests of wheat flour by Singh *et al.* (1993). The development of larvae of *Tribolium castaneum* was inhibited. The growth and development of first instar larvae of *Trogonoderma granarium* were completely arrested. It was effective against *Corcyra cephalonica* and *Sitotroga cerealella*. Chau (1993) also found neem preparations effective against insects of stored mung bean and rice and for the control of confused flour beetle (*Tribolium confusum*). Pandao *et al.* (1993) observed the efficacy of 5 percent neem extract against *Exelastis atomosa* in arhar (*Cajanus cajan* L.). Trung *et al.* (1993) described the development of neem pesticide for storage in Vietnam and Hu and Chiu (1993) in south China.

Belko (1994) mentioned that small farmers in Nigeria traditionally use neem leaves to control insects during the storage of cow pea. Neem oil at 1 percent concentration showed a significant reduction of egg hatching by the pulse beetle *C. chinensis* (Kachare *et al.*, 1994). The toxic effects of acetone and methanol extracts of whole neem fruit were tested by Majeed *et al.* (1994) on the pulse beetle *C. analis*, and these extracts were found to be quite effective, while Juneja and Patel (1994) treated green gram (*Vigna radiata*) with seed kernel powder against *C. analis* but it gave protection for three months only. Raju *et al.* (1994) observed the mortality of *Plutella xylostella* by commercial-grade neem oil, while Tabassum *et al.* (1994) tried neem preparations against *C. analis* and Naqvi *et al.* (1994) tried neem oil against it, and found them effective in controlling this pest.

Borker and Pawar (1995) found 1 percent neem seed powder to be a grain protectant against *C. chinensis*. Xie *et al.* (1995) tested three products containing different concentrations of azadirachtin and neem extract against three stored product insects, *S. oryzae*, *T. castaneum* and *Cryptolestes ferrugineus*. The neem extract was more effective than azadirachtin, which showed that the activity was not due to azadirachtin only but because of other compounds in the neem also. In the case of azadirachtin, the activity was dose dependent, i.e. the higher the concentration, the greater the effect. Sharma (1995) observed that 10 percent neem seed kernel powder was effective in stored maize for all the insects but did not provide complete protection. Niber (1995) studied the protective effect of stored maize against *Prostephanous truncatus* Gangopadhyay (1995) studied the use of turmeric (*Curcuma longa*) and neem for safe storage of food grains. Singh *et al.* (1996, 1996a) found that extract of neem was effective against the lesser grain borer *R. dominica* in terms of lower fecundity, adult mortality and less grain damage.

TREATMENT OF STORAGE AREA

A herbal formulation from neem, *Pongamia pinnata* and *Vitex negundo* was tried by Dakshinamurthy (1993) for the storage of wheat and pulses free from worms like *S. oryzae*, *C. maculatus* and *Corcyra cephalonica*, by impregnation of bags and surface treatment of storage bins. With this treatment, the level of infestation could be reduced considerably. Rajesh Kumar *et al.* (1994) observed the effect of soaking bags in a commercial neem preparation for the storage of rice. There was some protection.

PRESERVATION OF FISH

Okorie *et al.* (1991) placed Tilapia fish with neem seed powder and the insect *Dermestes maculatus*. The neem seed inhibited oviposition in the insect and killed the adults; most of the larvae in the fish did not develop and died within 30 days. Ward and Golob (1994) discussed various plant materials, including neem, to control insect infestation of cured fish.

FUNGISTATIC ACTION

Dry Seed

Seed stored in hot, humid and dark conditions often have fungal growth, sometimes consisting of *Aspergillus flavus* and *A. parasiticus*. These fungi produce a very highly carcinogenic substance, aflatoxin. An aflatoxin content exceeding 20 parts per billion (ppb) in animal feed and 0.5 ppb in eggs and milk is not permitted in the United States. Neem seed very often harbour these fungi when fresh but neem leaf extract has been found to suppress the production of aflatoxin in groundnut (Gherwande and Nagaraj, 1987). Zeringue and Bhatnagar (1990) observed that in cotton bolls, fungal growth was unaffected by aqueous extract of neem but it caused 16 percent inhibition of aflatoxin production. Bansal and Sobti (1990) controlled the growth of *Aspergillus* spp. by soaking peanut seed in neem extract. Khan and Shah (1992) studied the antifungal activity of leaf extract of neem on seed mycoflora of neem. Fresh leaves blended with potassium phosphate solution, when added to the fungal growth medium, did not affect fungal growth but stopped the synthesis of aflatoxins (Anonymous, 1993). *In vitro* studies with fungi suggested that non-volatile neem leaf constituents inhibit aflatoxin biosynthesis in the early stages of biosynthetic pathways (Bhatnagar and Zeringue, 1993). Up to 92 percent suppression of aflatoxins by methanolic extract has been observed by Shankar Rao *et al.* (1994). There was no correlation between growth of *A. flavus* and aflatoxin production.

Fresh Fruit

Fresh fruits and vegetables are often spoiled during storage and transit, particularly in hot and humid conditions, due to fungal growth. Some studies have been conducted to see if the shelf life of these products can be prolonged by treatment with neem products. Hasabnis and D'Souza (1987) studied post-harvest storage by dipping Alphonso mango fruits in neem leaf extract and by lining bamboo packing baskets with neem leaves as cushioning material. Ali *et al.* (1992) evaluated neem oil, leaf extract and pericarp dust against isolates of *Penicillium italicum*, *Alternaria alternata* and *Aspergillus niger* from rotten fruit. Neem oil was as effective as thiabendazole in checking growth of these fungi in rotting tomatoes. Moline and Locke (1993) tested the antifungal properties of a hydrophobic seed extract against post-harvest mango and apple pathogens, *Botrytis cinerea*, *Penicillium expansum*, and *Glomerella cingulata*. Neem seed was as effective as calcium chloride. Apples dipped in 2 percent emulsion of a commercial product after harvesting and stored at 0°C for four months had 30 percent less decay than those kept at room temperature (Hohn *et al.*, 1996).

REFERENCES

Anonymous (1993) Neem reduces aflatoxin levels. *Neem Newsletter*, **1**, 4–5.
Ali, T.E.S., Nasir, M.A. and Shakir, A.S. (1992) *In vitro* evaluation of certain neem products as mould inhibitors against post-harvest rotting fungi of tomato. *Pakistan Journal of Phytopathology*, **4**, 61–68.
Bansal, A.R.K. and Sobti, A.K. (1990) An economic control of two species of Aspergillus on groundnut. *Indian Phytopathology*, **43**, 451–452.
Belko, H. (1994) Efficacy of traditional methods of storage of cowpeas in the rural development of Niger. *Sahel PV-Information* No. **68**, 2–8.
Bhatnagar, D. and Zeringue, H.J. Jr. (1993) Neem leaf extracts (Azadirachta indica) inhibit aflatoxin biosynthesis in Aspergillus flavus and A. parasiticus (abstr.). *Proc. World Neem Conf.*, 24–28 Feb. 1993, Bangalore, India.

Borker, P.S. and Pawar, V.M. (1995) Relative efficacy of some grain protectants against Callosobruchus chinensis (Linnaeus). *Pesticide Research Journal*, **7**, 125–127.

Chau, L.M. (1993) Effect of Neem leaves and other botanicals to control rice plant hoppers, rice nematode and stored grain insects in the Mekong Delta (abstr.). *Proc. World Neem Conf.*, 24–28 Feb. 1993, Bangalore, India.

Cobbinah, J.R. and Appiak-Kwartang, J. (1989) Effect of some neem products on stored maize weevil, Sitophilus zeamais. *Insect Science and its Application*, **10**, 89–92.

Dabire, C. (1992) Traditional methods of protection of stores of cowpeas in Burkina Faso. *Sahel PV Information*, **49**, 7–13.

Dakshinamurthy, A. (1993) Technology for producing insecticides of plant origin at rural level (abstr.) *Proc. World Neem Conf.*, 24–28 Feb. 1993, Bangalore, India.

Das, G.P. (1989) Effect of the duration of storing chickpea seeds treated with neem oil on the oviposition of the bruchid Callosobruchus chinensis Linn. (Bruchidae: Coleopetra). *Bangladesh Journal of Zoology*, **17**, 199–201.

Gangopadhyay, S. (1995) Use of Curcuma longa, Azadirachta indica for safe storage of food grains. *International Conference, Current Progress on Medical and Aromatic Plants Research*, Calcutta, 30 Dec. 1994 to 1 January 1995.

Gherwande, M.P. and Nagaraj, G. (1987) Prevention of aflatoxins contamination through some commercial chemical products and plant extract in groundnut. *Mycotoxin Research*, **3**, 19–24.

Gupta, H.S., Verma, J.P., Bareth, S.S. and Mathur, B.N. (1988) Evaluation of some non-edible oils as grain protectant, and their subsequent effect on germination. *Indian Journal of Entomology*, **50**, 147–150.

Hasabnis, S.N. and D'Souza, T.F. (1987) Use of natural plant products in the control of storage rot in Alphonso mango fruits. *Journal of Maharashtra Agricultural University*, **12**, 1205–1206.

Hohn, H., Hopli, H.U. and Graft, B. (1996) Qassia and Neem: exotic insecticides in fruit culture. *Obst-und-Weinbau*, **132**, 62–63.

Hu, M.F. and Chiu, S.F. (1993) Experiments on the effectiveness of some botanical insecticides in controlling the confused flour beetle *Tribolium confusum* storage. *Journal of South China Agricultural University*, **14**, 32–37.

Ivbijaro, M.F. (1983) Toxicity of neem seed, Azadirachta indica A. Juss to Sitophilus oryzae (L) in stored maize. *Protection Ecology*, **5**, 353–357.

Islam, B.N. (1993) Notes on the traditional use of indigenous plants and weeds for pest control in Bangladesh (abstr.). *Proc. World Neem Conf.*, 24–28 Feb. 1983, Bangalore, India.

Jadhav, K.B. and Jadhav, L.D. (1984) Use of some vegetable oils, plant extracts, and synthetic products as protectants for pulse beetle, Callosobruchus maculatus Fabr. in stored gram. *Journal of Food Science and Technology, India*, **21**, 110–113.

Jilani, G. and Saxena, R.C. (1990) Repellent and feeding deterrent effects of turmeric oil, sweet flag oil, neem oil and a neem based insecticide against lesser grain borer (Coleopetra: Botrychidae). *Journal of Economic Entomology*, **83**, 629–634.

Juneja, R.P. and Patel, J.R. (1994) Botanical materials as protectant of green gram, Vigna radiata (L) against Pulse beetle, Callosobruchus analis fabricus-I. *Gujrat Agricultural University Research Journal*, **20**, 84–87.

Kachare, B.V., Khaire, V.M. and Mote, U.N. (1994) Efficacy of different vegetable oils as seed treatment in increasing storage ability of pigeon pea seed against pulse beetle, Callosobruchus chinensis L. *Indian Journal of Entomology*, **56**, 58–62.

Kahare, S.N., Kahare, N.P., Harinkhere, J.P., Kandalakr, V.S. and Thakur, S.K. (1993) Exploration of herbal products as grain protectant against Callosobruchus chinensis. *Journal of Soils and Crops*, **3**, 33–36.

Khan, M.I. and Shah, N.H. (1992) Antifungal activity of neem on seed mycoflora of wheat. *Bioved*, **3**, 209–210.

Kossou, D.K. (1989) Evaluation of different products of neem Azadirachta indica A. Juss for the control of Sitophilus zeamais Motsch on stored maize. *Insect Science and its Application*, **10**, 365–372.

Kumar, R. and Mehta, J.C. (1993) Comparative efficacy of some plant materials against Sitotroga cerealella Oliv and Rhyzopertha dominica in stored milled rice. (abstr.) *Proc. World Neem Conference*, 24–28 Feb. 1993, Bangalore, India.

Majeed, I., Imtiaz, A., Naqvi, S.N.H., Khan, A.R., Tabassum, R. and Imran Qureshi (1994) Determination of toxicity of neem extract (Nfc and N-7) and Coopex 25 EC (permethrin—bioallelthrin) on pulse beetle Callosobruchus analis. *Proceedings of 14th Pakistan Congress of Zoology, University of Karachi*, 1–3 April 1994.

Makanjuola, W.A. (1989) Evaluation of extracts of neem (Azadirachta indica) for the control of some stored product pests. *Journal of Stored Products Research*, **25**, 231–237.

Malik, M.M., Naqvi, S.H. Mujtaba (1984) Some indigenous plants as repellant plants or antifeedant for stored grains. *Journal of Stored Product Research*, **20**, 41–44.

Mohiuddin, S., Qureshi, R.A., Ahmed, Z., Qureshi, S.A., Jamil, K., Jyoti, K.N. and Prasuna, A.L.(1993) Laboratory evaluation of some vegetable oils as protectant of stored products. *Pakistan Journal of Scientific and Industrial Research*, **36**, 377–379.

Moline, H.E. and Locke, O.C. (1993) Comparing neem seed oil with calcium chloride and fungicides for controlling post-harvest apple decay. *Horticulture Science*, **28**, 710–720.

Niber, B.T. (1995) The protectant and toxicity of four plant species on stored maize against Prostephanous truncatus (Horn) Coleopetra: Bostrichidea. *Tropical Science*, **35**, 371–375.

Naqvi, S.N.H., Rukhsana, Gule Khan, M.Z., Azmi, M.A. and Rani, S. (1994) Effect of neem oil, deltamethrin and perfekthion on total esterases activity of Callosobruchus analis. *Proceedings of the 14th Pakistan Congress of Zoology, University of Karachi*, 1–3 April 1994.

Okorie, T.G., Siyanbola, O.G. and Ebochuo, V.O. (1991) Neem seed powder as a protectant for dried Tilapia fish against Dermestes maculatus Degeer infestation. *Review of Agricultural Entomology*, 790–793.

Pandao, S.K., Mahajan, K.R., Muqueem, A., Aherkar, S.K. and Thakre, H.S. (1993) Efficacy of some insecticide against tur pod borers on semi-rabi arhar (Cajanus cajan) var. C-11. *PKV Research Journal*, **17**, 229–230.

Pandey, N.D., Mathur, K.K., Pandey, S. and Tripathi, R.A. (1986) Effects of some plant extract against pulse beetle Callosobruchus chinensis Linnaeus. *Indian Journal of Entomology*, **48**, 85–90.

Rajesh Kumar, Mahla, J.C. and Vinod Kumar (1994) Effect of gunny bag treatment with insecticides and plant extracts. *Annals of Biology, Ludhiana, India*, **45**, 55–58.

Raju, S.V.S., Chaudhary, M.K. and Singh, H.N. (1994) Bio-efficacy of some commonly used insecticide against Plutella xylostella L. *Indian Journal of Entomology*, **56**, 246–250.

Shankar Rao, C., Elliah, P., Reddy, D.S., Krishnappa, K. and Prabhakar, G. (1994) Effect of methanolic extracts of plants on aflatoxin production. *Indian Journal of Natural Products*, **10**, 13–15.

Sharma, R.K. (1995) Suppression of insect reproductive potential by neem in stored maize. *Annals of Plant Protection Science*, **3**, 113–114.

Singh, A.K., Khan, A.M., Jain, M.K., Chandel, B.S. and Pandey, U.K. (1987) Insecticidal properties of some oils against Rizopertha dominica Faor. Bostrichidae: Coleopetra, a pest of stored grains. *Zeitschrift fur Angewandte Zoologie*, **74**, 411–415.

Singh, H., Mrig, K.K. and Mahla J.C. (1996) Effect of different plant products on the fecundity and emergence of lesser grain borer, Rhyzopertha dominica (F.) in wheat grains. *Annals of Biology, Ludhiana, India*, **12**, 96–98.

Singh, H., Mrig, K.K. and Mahla, J.C. (1996a) Efficacy and persistence of plant products against lesser grain borer, Rhyzopertha dominica (F.) in wheat grains. *Annals of Biology, Ludhiana, India*, **12**, 99–103.

Singh, R.P., Jhansi Rani and Doharey, B. (1993) Neem dust (Azadirachta indica A. Juss) for management of stored grain pests. *Proc. World Neem Conference*, Banglore, India, 24–28 Feb. 1993.

Singh, R.P. and Kataria, P.K. (1986) Deoiled neem kernel powder as protectant of wheat seed against Trogoderma granarium. *Indian Journal of Entomology*, **48**, 119–120.

Tabassum, R., Naqvi, S.N.H., Ahmad, J.V., Shaista Rani, Jahan, M. and Azami, M.A. (1994) Toxicity determination of different plant extracts (saponin and juliflorine) and neem based pesticide Margosan OTM against stored grain pest Callosobruchus analis. *Proceedings of 14th Congress of Zoology held at the University of Karachi, Pakistan*, 1–3 April, 1994.

Trung, La Minh, Trang, Ngyen and Duy Trang (1993) Present status of Research and development of botanical pesticides in Vietnam. (abstr.) *Proc. World Neem Conference*, Bangalore, India, 24–28 Feb. 1993.

Vijayalakshmi, K., Radha, K.S. and Vandana, S. (1995) *Neem: a User's Manual*. Research Foundation for Science and Technology, New Delhi, India.

Ward, A.R. and Golob, P. (1994) The use of plant materials to control insect infestation of cured fish. *Tropical Science*, **34**, 401–408.

Wealth of India (1948) Vol. I, Publication Division, CSIR, New Delhi.

Xie, T.S., Fields, P.G. and Isman, M.B. (1995) Repellency and toxicity of azadirachtin and neem concentrates to three stored products beetles. *Journal of Economic Entomology*, 88, 1024–1031.

Zhang, X. and Zhao, S.H. (1983) Experiments on some substances from plants for the control of rice weevils. *Journal of Grain Storage Klangshi Chucang*, **1**, 1–8.

Zeringue, H.J. Jr. and Bhatnagar, D. (1990) Inhibition of aflatoxin production in Aspergillus flavus infected cotton bolls after treatment with neem (Azadirachta indica) leaf extracts. *Journal of the Americal Oil Chemists Society*, **67**, 215–216.

20. COMPOSITE PLANT FOR UTILIZATION OF NEEM

Neem fruit can be a very good raw material for a composite plant for the manufacture of various industrial and consumer products as given in Fig. 17. The properly dried decorticated seed will yield the seed coat and the kernel.

THE SEED COAT (HUSK)

It is very rich in cellulose, lignin, etc. and can possibly be put to the following uses.

(a) *As a fuel*: It has a low calorific value and does not burn easily, and is fed into the furnace of the boiler with a spade so that it does not form a lump but gets spread on the flame for easy combustibility. It can be used in places where slow heat for a long time is required, as in brick kilns.

(b) *Briquettes*: After powdering (Fig. 18A) it may be mixed in a mixing machine with other unwanted organic materials of the industry like discarded neem oil, seed cake or other agricultural wastes and pressed to form briquettes with the help of a machine (Fig. 18B). These briquettes can be used in the boiler for the steam required for heating purposes.

(c) *The ash*: Seed coat is very rich in inorganic matter, and yields a substantial amount of ash, which if not disposed of properly can not only create storage problem but may cause pollution by increasing the amount of suspended particles in the air. The best use of it can be as an ingredient in cement industry in place of fly ash obtained from thermal electric plants. It may be incorporated in clay for the brick industry.

(d) *Particle Board*: The seed coat has good mechanical strength and may be incorporated in the mixture of wood shavings used for making particle boards for thermal insulation by mixing with a synthetic resin, such as phenol formaldehyde, as adhesive. It may be powdered before use or treated with acid or alkali so that it forms a homogenous mass with the particle board mixture.

(e) *Manure*: It can be incorporated in soil amendment formulations, particularly for soil rich in clay or having nematodes. In the former case, it loosens the soil particles while in the latter case it may destroy nematodes by the traces of limonoids contained in it.

(f) *As a raw material*: Some new methods can be developed for other industrial products after chemical treatment.

SEED OIL

Earlier, neem oil on an industrial scale in India was obtained mainly by solvent extraction, but with the demand for pesticides, modifications to the technology have been suggested so that azadirachtin and allied compounds can also be recovered. Depending

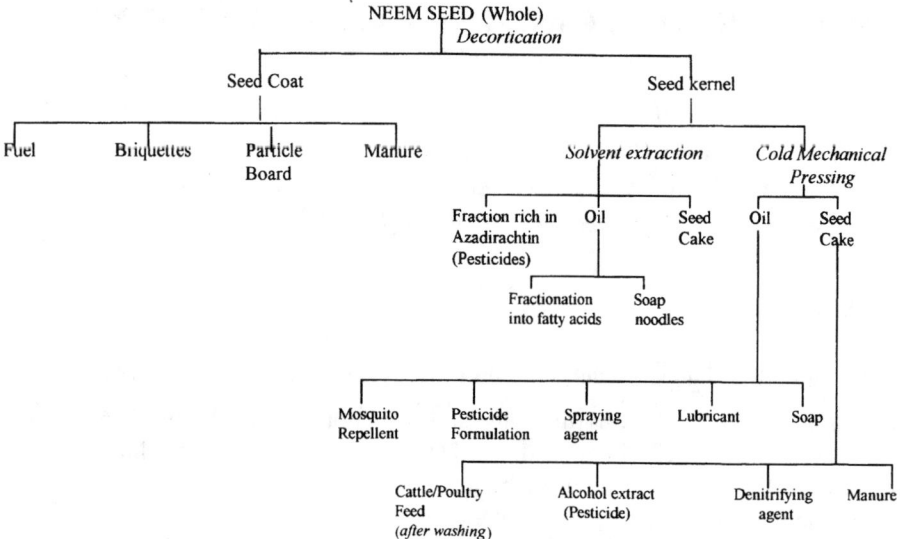

Figure 17 Flow diagram for the utilization of neem for the manufacture of various products

on the requirement, the oil from seed kernel powder can be extracted either by solvents or by cold mechanical pressing.

By Solvents

As given in the process for the isolation of azadirachtin in Chapter 3 on chemical constituents, the seed is powdered to a specific particle size. If it is coarse, proper extraction will not take place; on the other hand, fine powder may clog the pipes and filtration of oil may be difficult. When seed is extracted with solvents, limonoids, other constituents and the oil get dissolved in it, leaving the seed cake, free of these, in the extractor. The solvent from this mixture can be recovered by distillation. For pesticide formulations, pesticides are removed from the oil by using other solvents, and standardized extract may be used in the desired products.

The fatty oil obtained may be used in the manufacture of soap by mixing with other oils, but for good quality soap, noodles are preferred which can be manufactured by treating the oil with sodium hydroxide. Only the fatty acids react with caustic soda, leaving most of the impurities, which affect the quality of the soap. These soap noodles are mixed with various ingredients like silicates, polyethylene glycols, surfactants, color and fragrances, and kneaded together. The mixture is fed into an extruder to get the soap cakes. The oil can be split further by various chemical treatments to get free fatty acids, which are very important raw materials in the detergent, cosmetic and rubber industries. Glycerine is an important by-product. The fatty acids may be converted into lubricants after hydrogenation or may be used for the manufacture of olein and high melting styrene.

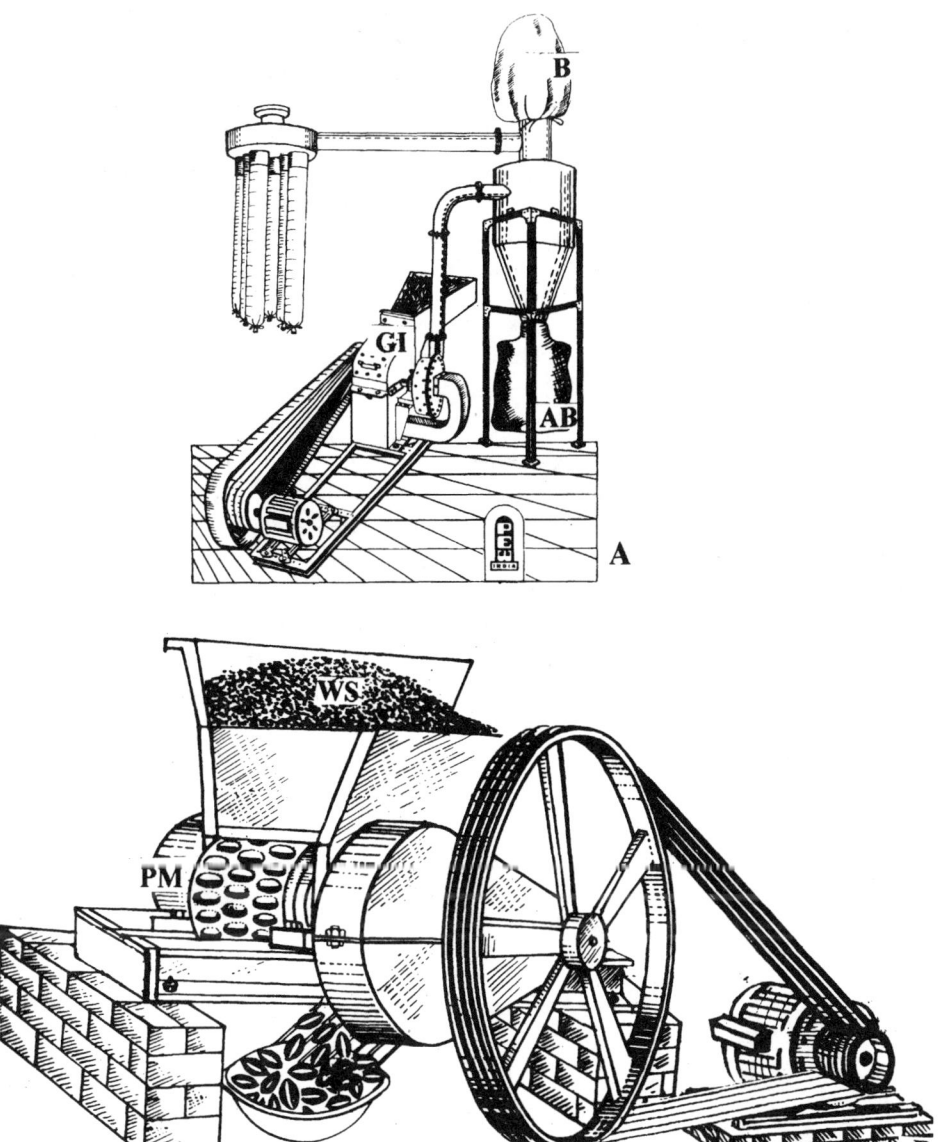

Figure 18 A, Machine for making fine powder of neem seed coat and other agricultural waste; B, briquette-making machine. *Abbreviations*: AB = air balloon, GI = grinder, WS = waste stuff, PM = pressing machine

Cold mechanical process

In this, the seed kernels are subjected to mechanical pressing, and no solvents, chemicals or heat is used. By this process, oil rich in limonoids is obtained but the yield is low. This oil may be further enriched by azadirachtin and the other compounds obtained from solvent extraction for pesticide formulations or may be used as such or in the form of an emulsion as a spraying agent or in mosquito-repellent products, by incorporating it in other oils or in a cream or paste base. The soap manufactured from this oil can be effective in skin diseases. The oil may be used as a lubricant for household and agricultural purposes.

SEED CAKE

The cake obtained after solvent extraction has a small amount of oil and limonoids and after washing with water can be incorporated into poultry and cattle feed, as given earlier, but the cake obtained from the cold process is rich both in oil and limonoids and may be extracted with ethanol for use in pesticide formulations and the extracted seed cake may be used as a denitrifying agent or as a manure. For a nematode suppressant formulation, neem cake mixed with karanj (*Pongamia glabra*) seed cake and tobacco waste may be used. The seed cake may be mixed with urea for denitrifying purposes in fertilizers. Neem cake manure has better customer acceptance for household purposes and indoor plants because it lacks the malodor of cattle manure and has good nutritive value for the vegetation.

21. PATENTS ON NEEM

EARLIER PATENTS

As given in the list of patents in Appendix 1 of this chapter, Indian workers were the first to isolate neem bitters in 1945–46 and they got patents for these, but the patents could not be effectively utilized by the industry because no product which might have good demand could be developed from these bitters. Later on, the insecticidal value of neem was observed by Indian scientists but they were not granted a patent for their discovery, because of the common use and common knowledge clause in the Indian Patent Act. The Indian Central Insecticides Board did not register neem products under the Insecticide Act 1968 (Vijayalakshmi *et al.*, 1995). In England and Japan, a few bodycare products with neem as an ingredient were patented.

NEEM PATENT CONTROVERSY

The controversy started when a patent was granted to W.R. Grace and Co. for a neem-based pesticide. The question arose whether a product of natural origin, the traditional use of which has been known since ancient times, can be patented. More than 200 organizations from 35 countries raised their voice, against this type of patent, which was called "Corporate Colonialism", "Genetic Imperialism" and "Folk-wisdom Piracy". Vijayalakshmi *et al.* (1995) have given details of these activities, as shown in the poster (Fig. 19).

W.R. Grace and Co. defended their patent right under the pretext that they had developed a process for the isolation of a stable form of azadirachtin and for increasing the shelf life of the product. In 1993, a Congressional Research Service (CRS) reported to the United States Congress put forth the view that synthetic form of natural products like azadirachtin may be patentable (Vijayalakshmi *et al.*, 1995).

In an editorial in the magazine *Nature* (Anonymous, 1995), patents on neem were justified on the ground that companies have to spend a huge amount of money on research and development for this type of product, and have to protect this investment by taking out patents on their discoveries. The author agreed that in this way, general knowledge becomes a private commodity.

In reply to this, Balasubramanian (1995) under patents and indigenous lore mentioned that patents for natural products should be granted on their genuine originality and not to the extent they conflict with traditional knowledge systems, turning public goods into a private commodity. The author feared that large-scale purchase of the raw materials by multinationals, which have huge resources at their disposal, may take the price of unprocessed neem beyond the reach of farmers who may be forced to rely on the commercial product rather than on traditional recipes.

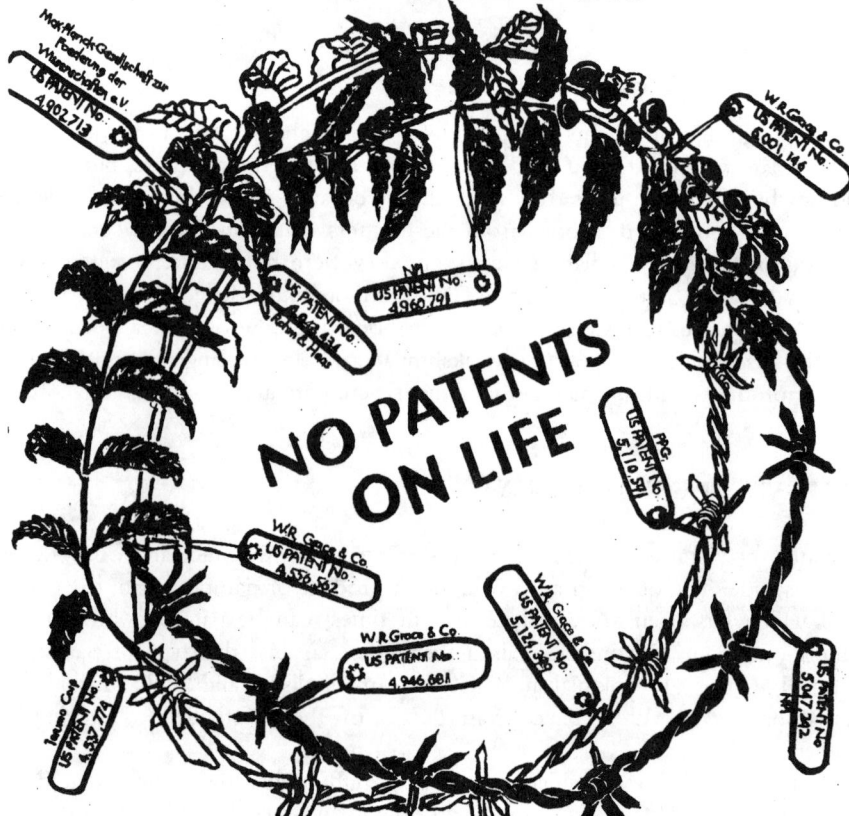

Figure 19 Poster against granting patent rights to neem products

The role played by advanced technology in products developed from natural resources like neem have been discussed by Hoyle (1995). The author quoted an instance of India's own Central Drug Research Institute which has transferred the technology for natural spermicide DK-1 to two companies. Crsepi (1995) suggested ways by which the patents on products developed by biotechnology can be defended.

The whole issue of patents for neem products has been discussed by Johnston (1995). According to the author, the Neem campaign feels that the utilization of traditional knowledge, which is an intellectual contribution of third world countries, for commercial gain, is a sophisticated name for modern piracy. The Foundation on Economic Trends in Washington DC requested the US Patent and Trademark office to overturn patents on neem and sought legislation to bar this type of patent in future (Anonymous, 1996).

APPENDIX 1

Patents on Neem (in Chronological Order)

Siddiqui, S. and Mitra, C. (1945) Isolation of new bitter principles from neem oil. Indian Patent No. **33640**.

Siddiqui, S. and Mitra, C. (1945a) Separation of the bitter principles of neem oil and simultaneous refining of oil. Indian Patent No. **33650**.

Siddiqui, S. and Mitra, C. (1945b) Manufacture of new derivatives or subsidiary products from the physiologically active bitter principles of neem oil. Indian Patent No. **33651**.

Bhattacharji, S. and Mitra, C.R. (1953) Isolation and separation of bitter principles from the trunk or root bark of nim. Indian Patent No. **46947**.

Mitra, C. (1953) Process for the thermal degradation of the seed oil of nim (Melia azadirachta). Indian Patent No. **48529**.

Thakur, M.S. and Godrej, B.P.C. (1972) Purifying neem oil. Indian Patent No. **118768**.

Sawanobori, H., Tanaka, H., Saito, K., Tekeuchi, Y., Shirasawa, K. and Saito, S. (1977) Melia azadirachta extracts for skin cosmetics. Japan Patent Nos. **7728853** and **7728854**.

Latiff, A. (1979) Formulation for treating hair. British UK Patent, **2000971**.

Masaki, Shimizu, Tadashi, Sudo and Tokeo Nomura (Terumo Corporation) (1985) Neem bark extracts, US Patent No. **4515785**.

Masaki, Shimizu, Tadashi, Sudo and Tokeo Nomura (Terumo Corporation) (1985) Hot water extracts of Neem. US Patent No. **4537774**.

Robert, O., Larson (Vikwood Ltd., now taken over by W.R. Grace & Co.) (1985) Stable neem seed anti-pest neem seed extract. US Patent No. **4556562**.

Rembold, H. et al. (Max-Planck Gesellschaft Zur Foederung der Wissenschaften e.V., Germany) (1990) Azadirachtin like compounds and insect destroying agents containing them. US Patent No. **4902713**.

Lidert, Z. (Rohm & Haas Company) (1990) Insecticidal hydrogenated neem extracts. US Patent No. **4943434**.

James, F. Walter (W.R. Grace & Co.) (1990) Methods to prepare an improved storage stable neem seed extract. US Patent No. **4946681**.

James, A. Klocke et al. (Native Plant Institute) (1990) Salannin derivative insect control agents. US Patent No. **4960791**.

Charles, G., Carter et al. (W.R. Grace & Co.) (1991) Storage stable azadirachtin formulation. US Patent No. **5001146**.

James, A. Klocke et al. (Native Plant Institute) (1991) Azadirachtin derivative insecticide. US Patent No. **5001149**.

PPG Inc. (1991) Concentrated matter in oil microcmulsion that forms storage stable oil in water emulsion consisting of 50% to 90% by weight of neem oil containing the pesticide azadirachtin, among other ingredients. US Patent No. **5110591**.

Floss Products Corp. (1991) US Patent No. **5009886**.

PPG Inc. (1992) Neem Oil Emulsifier. US Patent No. **5110491**.

W.R. Grace & Co. (1992) Storage stable azadirachtin formulation. US Patent No. **512449**.

Godrej Soaps (India) (1994) Neem oil fatty acid distillation residue based pesticide. US Patent No. **5298247**.

Nature Plants (1994) Chemical pure compound deemed from naturally insect repellents. US Patent No. **5247242**.

W.R. Grace & Co. (1994) Storage stable pesticides composition contains neem seed extract solution, US Patent No. **5281618**.

W.R. Grace & Co. (1994) Fungicide compositions derived from neem oil and neem wax formulation. US Patent No. **5298251**.

FMC Philadelphia (1994) Ascaricidal combination of neem seed extract and bifenthrin. US Patent No. **5352697**.

W.R. Grace and US Department of Agriculture (1994) Hydrophobic extracted neem oil, a novel fungicide. US Patent No. **5356628**.

James Charles Locke, Hiram Gordon Larew III (James Frederic Walter) (1994) Method for controlling fungi on plants by the aid of hydrophobic extracted neem oil. European Patent No. **0436257 B1**.

Lidert, Z. (Rohm & Hass Company) (1995) Edible neem oil. US Patent No. **5371254**.

REFERENCES

Anonymous (1995) Patenting nature now. *Nature—London*, **377**, 89–90.
Anonymous (1996) Neem patent Controversy. *American Herb Association Quarterly Newsletter*, **12**(3), 8.
Balasubramanian, D. (1995) Patents and indigenous lore. *Nature—London*, **378**, 352.
Crepsi, S. (1995) Biotechnology patenting: the wicked animal must defend itself. *European Intellectual Property Review*, **17**, 431–444.
Hoyle, R. (1995) Neem underscores need for US biodiversity role. *Bio-Technology*, **13**, 1164.
Johnston, B. (1995) The patenting of neem: modern piracy or political correctness? *Herbalgram*, **35**, 32–33.
Vijayalakshmi, K., Radha, K.S. and Vandana, S. (1995) *Neem—A User's Manual*. Research Foundation for Science, Technology and Natural Resource Policy, New Delhi.

SUBJECT INDEX

$1\beta,2$-diepoxyazadirdrone, 95
5'nucleotidase, 115
6-benzoylamino purine, 39
7-acetylneo-trichlenone, 95
β cells, 88
β sitosterol, 27, 38
1 interferon, 98

Abortifacient, 112
Acacia, 42
Acacia nilotica, 63
Acacia tortilis, 42
Acanthocheilonema vitae, 87
Acetylcholine, 95
Acetylsalicylic acid, 92
Acroparesthesia, 84
Actinomycetes, 130
Acute articular rheumatism, 84
Adaptogenic effect, 98
Adhatoda vasica, 78
Adrenaline, 88, 92
Aedes aegypti, 147
Aerobic bacteria, 130
Afforestation, 33, 42, 61
Aflatoxicosis, 74
Aflatoxin, 35, 116, 151, 154
Africa, 41, 125
Agrobacterium tumefaciens, 40
Air freshner, 141
Air layering, 37
Air purifier, 141
Albizzia lebbeck, 42
Alcohol segregation, 70
Aldobiuronic acid, 27
Aldotriuronic acid, 27
Alkaline phosphate, 97
Alphonso mango, 154
Alternaria spp., 45
Alternaria alternata, 154
Alum, 79
Amargosiera, 11
Amblyomma americanum, 148

Ambrosia, 1
Ames test, 117
Amino acid, 27, 74, 117, 135
Aminobutyric acid, 27
Ammoniated nitrogen, 130
Amnesia, 84
Amrita, 1
Anal itching, 82
Anal region, 87
Analgesic, 93, 96
Andhra Pradesh, 9
Anemophilous, 16
Ankylostomiasis, 87
Annual rings, 61
Anomocytic, 58
Anona reticulata, 122
Anopheles stephensi, 146, 147
Anopheline pupae, 147
Ant, 145
Anthelmintic, 82, 87
Anthesis, 25
Antiandrogenic, 90, 91
Antianxiety, 97
Antibacterial, 87, 88
Antibacterial effect, 130
Anticancerous, 94
Anticholinergic, 96
Anticomplement, 97
Anticonvulsant, 96
Antidiabetic, 88
Antiestrogenic, 90
Antifeedant, 87, 122, 123, 148
Antifertility, 89, 90, 112
Antifilarial, 87
Antiflagellate activity, 89
Antifungal, 91
Antigastric, 95
Antihistamine, 97
Antiimplantation, 89
Antiinflammatory, 79, 80, 83, 92, 93, 113
Antileprotic, 99

Antilice, 113
Antimalarial, 78, 94
Antimicrobial, 79, 111
Antineoplastic action, 95
Antinociceptive, 93
Antiovulatory, 89
Antipeptic ulcer, 95
Antiproteolytic, 99
Antiprotozoal, 93
Antipyretic, 92
Antirheumatoid, 93
Antiseptic, 82, 113
Antiseptic agent, 81
Antiseptic cream, 111
Antispermatic, 91
Antistress, 96, 97
Antitubercular, 88
Antitumor, 94
Antiviral, 95, 96, 122
Anxiolytic, 97
Aparjith Dhup, 78
Aphids, 72
Apocolpium, 15
Applied entomology, 125
Aquatic life, 123
Aquic petrocalcic natrustalf, 43
Arabinose, 27, 97
Archidic acid, 59
Archidonic acid, 134
Argenine, 27, 135
Arhar, 152
Arishta, 77
Arq Gaz, 83
Arq harabhara, 83
Arq murakkab musaffa khun, 83
Arsenical dermatitis, 98
Arterial blood, 96
Arthritic, 92
Arthropods, 45, 122
Asav, 77
Ascaricidal, 166
Ascaridia galli, 87
Ascaris oxyrus, 87
Ascaris lumbricoides, 87
Ascaris vermicularis, 87

Ascites tumor, 95
Aspartic acid, 135
Aspergillus flavus, 153, 154
Aspergillus fumigatus, 45
Aspergillus niger, 45, 154
Aspergillus parasiticus, 153
Astringent, 79
Ataxia, 134
Australia, 41
Autopsy, 91
Avian malaria, 93
Axial parenchyma, 61
Ayurveda, 79, 87, 113, 115
Ayurvedic preparations, 77, 97, 98, 99
Ayurvedic Pharmacopoeia, 78
Azadirachta excelsa, 17
Azadirachta indica var. *siamensis*, 17
Azadirachta integrifolia, 17
Azadirachta integrifoliola, 17
Azadirachta siamensis, 17
Azadirachtin, 94, 95, 145, 146, 147, 153, 149, 160, 163, 165, 166
Azadirone, 27

Bacillus anthracis, 88
Bacillus mycoides, 88
Bacillus subtilis, 88
Bacteria, 125, 130, 188
Bacterial infestation, 75
Bacteriostatic, 129
Baffles, 55
Baldness, 113
Bangladesh, 151
Bark, 59, 79
Bed bug, 145
Benzoyl benzoate, 98
Benzyladenine, 39
Bifenthrin, 166
Bilirubin, 91, 116
Bioefficacy, 123
Bioregulators, 130
Birds, 145
Biting louse, 111
Bitter, 87, 88, 134, 146, 147
Bitter principle, 69, 79, 133

Black pepper, 80
Blastocidal, 90
Blastocyte, 90
Bleeding gums, 79
Bleeding piles, 82
Blood brain barrier, 97, 124
Blood cells, 136
Blood glucose concentration, 88
Blood purifier, 79, 82, 83
Blood urea, 116
Bloodless piles, 83
Blower, 67
Blue green algae, 129
Bortytis cinerea, 154
Bovicola ovis, 111
Briquettes, 159
Buffalo, 136
Burma, 75
Butea frondosa, 152
Butter oil, 80

Calcium, 43, 135, 136
Calcium activation, 93
Calcium chloride, 42, 154
California, 41
Calipitrimerus azadirachtae, 46
Callosobruchus analis, 153
Callosobruchus chinensis, 152, 153
Callosobruchus maculatus, 152, 153
Callus, 38, 39
Callus proliferation, 38
Calorific value, 62, 63, 159
Cambiform, 61
Candida albicans, 99
Capillary permeability, 100
Caput didymus, 91
Caragenin, 92
Carak samhita, 77
Carbohydrate, 135
Carbolic acid, 98
Carbon black, 113
Carbon dioxide, 44, 141
Carbon monoxide, 141
Carcinogenic, 154
Cardioactivity, 96

Cardiogenic glycosides, 94
Cardiovascular activity, 96
Caribbean, 41
Carica papaya, 113
Carotenoid, 143
Cassia fistula, 79
Cassia siamea, 42
Castor oil, 87
Cataract, 81
Catchment areas, 43
Catechin, 93
Cathartic, 87
Cattle feed, 133
Cauda didymus, 91
Cauda epididymus, 90
Caustic flakes, 73
Cedar wood oil, 111
Cedrus deodara, 111
Cellular defense, 97
Cellulose, 159
Cephalosporium, 45
Cercospora blight, 46
Cercospora leucostricta, 45, 46
Cercospora subsessilis, 45
Cerebral oedema, 117
Cervico-vaginal, 99
Cervix, 84
Chad, 41, 45
Chagas' disease, 94
Chawlmogra oil, 98
Chemiluminescence, 93, 97
Chemotaxonomic, 18
Chiasmata index, 17
Chick pea, 152, 153
Chikungunya virus, 96
China, 152
China berry, 121
Chitin synthesis, 122, 123
Chlamydia trachomatis, 99
Chloroplast, 142, 143
Chloroquin, 93, 94
Cholorophyll, 142, 143
Chromatography, 72, 74
Chromosome, 17
Churn, 77

SUBJECT INDEX

Chutney, 80, 81
Cimex lectularius, 145
Cinchona, 79, 93
Citric acid, 27
Cluicine mosquito, 147
CNS (Central Nervous System), 96, 97
Cockroach, 145
Coconut oil, 73
Coenzyme Q, 117
Colletotrichum capsici, 45
Colletotrichum gloeosporioides, 46
Coloring matter, 134
Commiphora wightii, 147
Compressive strength, 62
Congressional Research Service, 163
Conjuctivitis, 83
Connective tissue, 39
Constipation, 84
Contour ditches, 37
Contraceptive, 90, 91
Cooling agent, 79
Copper sulfate, 111
Corcyra cephalonica, 152, 153
Cork layers, 83
Corporate colonialism, 163
Corpus cardiacum, 124
Cortex, 18
Cortex, secondary, 58, 59
Corticium salmonicolor, 45
Coryllium, 113
Cosmetic, 160
Cotton seed oil, 73
Cotyledons, 38, 39
Cow pea, 152
Croton oil, 92
Crude fiber, 133
Crude protein, 133
Cryptolestes ferrugineus, 153
Cuba, 41
Culex quinquefasciatus, 146, 147
Culture maintenance, 39
Culture media, 38, 39
Culture, tissue, 40
Curcuma longa, 78, 153

Custard apple, 122
Cutaneous lesions, 111
Cuticle, 122, 123
Cyclas puncticollis, 152
Cysteine, 135
Cytoplasm, 142, 143
Cytotoxicity, 91

Dalbergia sisoo, 44
Dandruff, 82, 113
Debility, 82
Deciduous forest, 40
Decorticator, 66
Dehusking, 66
Delmethrin, 146
Demcodecosis in dogs, 111
Demcodectic mange, 111
Dementia, 84
Demodex canis, 111
Demulcent, 80
Dengue fevers, 146
Denitrifying, 162
Dental care, 81
Depressed state, 84
Dermataceous hyphomycetes, 46
Dermatophytes, 98, 99
Dermestes maculatus, 153
Detergent, 160
Dhanwantri, 1
Dharek, 10
Dhattur tailam, 78
Dhupan, 77
Diabetes, 80, 82
Diabetes mellitus, 88
Diarrhoea, 137
Dicotyledonous, 58
Diesel engine, 73
Digestible coefficient, 133
Digestible energy, 133
Dilated pupils, 115
Dimness of sight, 115
Disruptive activity, 123
Distillative deacidification, 70, 73
Diuretic, 96
Divine tree, 1

DNA, 95
Dogonyaro, 11
Domnican Republic, 41
Dopamine, 98
Dribbing, 42
Drowsiness, 116
Drupe, 56
Druses, 58
Duodenal lesions, 95
Duodenal ulcer, 95

Earthworm, 123, 129, 143
Ecdysone, 122
Eclipta alba, 82
Eco-balance, 33
Ectoparasites, 83
Ectoparasitic insects, 111
Eczema, 82, 98
Egypt, 41
Electrophoresis, 39
Embelia ribes, 148
Emblica officinalis, 80
Embryo, 39, 91
Embryo sac, 15
Embryogenic culture, 39
Embryogenesis, 39
Embryogenesis, somatic, 39, 40
Embryoid, 39
Emetic, 87
Emmenagogue, 83
Emollient, 113
Encephalitis, Japanese, 146, 147
Endocarp, 16, 17, 34, 35, 56
Endocrine system, 124
Endometrium, 90
Endosperm, 123
Entomophilous, 16
Enzyme indicies, 116
EPA (Environmental Protection Agency), 2
Epicarp, 16, 56
Epicatechins, 93, 97
Epidermal cells, 18, 39, 142
Epidermis, 58, 115
Epidermophyton flocossum, 92
Epididymus, 91

Epigallocatechin, 93, 97
Epinimbin, 130
Epithelial cells, 90
Epithelization, 100
Eryspelas, 98
Essential oil, 27, 88
Estrogenic activity, 89, 90
Ethephon, 19
Ether extract, 133
Ethiopia, 41
Ethno-botanical, 9
Etiology, 82
Eucalyptus camaldulensis, 42
Eucamptodemous, 14
Euphorbia, 37
European colonizers, 79
Exalbiuminous, 56
Exelastis atomosa, 152
Expeller, 65
Explants, 39, 40
Exserohilum spp., 46
Extra-floral nectaries, 12, 18
Extraction, 69
Extractives, 135
Exuviae, 122

Face fly, 146
Face packs, 113
Falcate, 13
Fasciola gigantica, 148
Fascioliasis, 148
Fatty acid, 20, 23, 160
Fecundity, 153
Feed, cattle, 4, 133, 135
Feed, poultry, 4, 133
Fiji, 41
Firang rog, 79
Fire ants, 45
Fire proofing, 62
First instar larvae, 147
Flagellate, 94
Flash Chromatography, 25
Flatulence, 84
Flavanol-o-glycosides, 97
Flavonoid, 3, 92

Flies, 151
Flocculation, 143
Floral initiation, 44
Florida, 41
Fly ash, 141, 142, 159
Foetal abnormality, 95, 115
Folk wisdom piracy, 163
Folklore, 87
Fomes, 45
Foreign disease, 79
Forest, dry deciduous, 40
Forests, tropical, 40
Forgetfulness, 84
Formalin, 92
Formica polyctena, 145
Free fatty acids, 134
French, 11
Fresh water snail, 123
Fructose, 27
Fruit shedding, 44
Fuel wood value index, 62
Fumaric acid, 27
Fumigation, 145
Fungal flora, 116
Fungal growth, 153, 154
Fungal hyphae, 46
Fungi, 123, 125, 130
Fungicidal activity, 4, 166
Fungistatic, 92
Furunculosis, 98
Fusarium, 45

Gallic acid, 97
Gallocatechins, 93, 97
Gametophyte, 15
Gandoerma root, 46
Gangetic plains, 42
Ganoderma applanatum, 45
Ganoderma lucidum, 46
Gardnerella vaginalis, 99
Garlic, 79
Garuda, 1
Gastric mucosa, 95
Gastric ulcer, 95
Gastroenteritis, 137

Gastrointestinal irritant, 115
Gedunin, 25, 27, 94
Generalized seizures, 116
Genetic diversity, 40
Genetic imperialism, 163
Genetic improvement, 39, 40
Genetic recombination, 17
Gentian, 80
Geremplasm, 40
German, 11
Germinability, 38
Germination capacity, 34
Ghana, 62
Ghritam, 77
Giddiness, 84, 121
Gingivitis, 137
Glabrous, 17
Glomerella cingulata, 45, 46, 154
Glomus, 11
Glucose, 27, 97
Glue adhesion, 62
Glutamic acid, 135
Glycerides, 59
Glycerine, 160
Glycine, 27, 38, 135
Glycolether, 72
Glycogen, 116
Glycosides, 94
Glycyrrhiza glabra, 80
Goat, 137
Godavri, 9
GOT (Glutamate oxaloacetae aminase), 97
Gout, 92
GPT (Glutamate pyruvate transaminase), 97
Granulocytes, 98
Granuloma, 92
Greco-Persian, 1, 82
Green gram, 153
Greying of hair, 113
Growth disruptive, 124
Growth inhibition, 123
Growth inhibitor, 145
Growth inhibitory activity, 24

SUBJECT INDEX

Growth regulator, 37
Gujarat, 3, 134
Glutamine, 27
Glutamate, 97
Gynecological practice, 80
Gynoecium, 15

Hab bawasir badi, 83
Hab musaffi khun, 83
Hab narkachur, 83
Hab Siyah Chashan, 83
Hainan, 41
Hair care, 82
Haiti, 41
Hard abscesses, 82
Hardwood, 38, 61
Headache, 82, 84
Heal all, 11
Heart problems, 82
Hebbevu, 11
Hellopeltis antonii, 46
Hemicellulose, 27
Hemoglobin, 94, 136
Hemorrhoids, 80, 83
Hendersonula toruloidea, 99
Hepatic dysfunction, 116
Hepatic enzymes, 117
Hepatic glycogen, 89
Hepatobiliary toxicity, 115, 116
Hepatonephropathy, 134
Hepatoprotective, 97
Hermaphrodite, 15
Hexatriterpenoids, 23
Himalaya, western, 9
Hindu mythology, 1
Histamin, 95
Histochemical, 111
Histological differentiation, 38
Histopathological examination, 111
Histopathological changes, 134
HIV antieffect, 98
Homeopathy, 83, 87
Homoeopathic medicine, 55
Honey bee, 123, 143
Hormonal regulation, 90, 117

Hormone, 38
Hotness, 79
House fly, 146
Household pests, 45
HPLC (High Performance Liquid Chromatography), 25, 71
Human papilloma virus, 96, 99
Husk, 159
Hydnocarpus kurzii, 98
Hydrogen peroxide, 74
Hydrogenation, 74, 160
Hydrophobic, 154, 166
Hyperglycemic, 88
Hyperkeratosis, 115
Hyperpyrexia, 92
Hypertension, 96
Hypochondria, 84
Hypocotyl, 17, 39
Hypogymous, 15

IAA (Indole acetic acid), 38
IBA (Indole butyric acid), 37, 38
IgE-reactive proteins, 143
Imbricate, 15
Immunocontraception, 90
Immunological, 98
Immunomodulater, 94, 97, 98
Immunostimulating, 99
Imparipinnate, 13
Impure blood, 79
Indian Homeopathic Pharmacopoeia, 59, 83
Indian lilac tree, 11
Indian traditions, 115
Indischer Zedrach, 11
Indo-Gangetic plains, 1, 43
Indonesia, 1, 11, 25, 40, 41
Indoplanorbis exustsus, 148
Indra, 1
Inflammation, 80
Inflammatory stomatis, 92
Inhibitory compounds, 34
Insect feeding deterrent, 122
Insecticidal, 122
Insomnia, 84
Insufficient bowels, 84

Insulin, 88
Integrated pest management, 123
Intorse, 15
Intran, 11
Isoleucine, 135
Isoprenylated flavone, 18
Itch, 82
Ivory coast, 41

Japanese encephalitis, 146, 147
Jatayadi tailam, 78
Jaundice, 9, 97
Java, 40, 41
Jeevantiadi kashyam, 78
Joint pains, 82
Juvenile hormone, 122

Kaempferol, 27
Kai-bevy, 11
Kajal, 113
Kakopholo, 11
Kamakha, 11
Kanad, 11
Kandavadu lepah, 78
Karanj, 82
Karnataka, 9
Katchi Ghani, 65, 67
Kenya, 41
Keratinophilic fungi, 91
Keratomycosis, 99
Kerosene oil, 146
Khand, 77
Kinetin, 39
KOH (Potassium hydroxide), 134
Kohlu, 65, 67
Kohumba, 11
Kols, 9
Konkan, 11
Krimighana, 77
Kwath, 77

Lactating buffaloes, 136
Lactose, 93
Laghu manijshtadi kwath, 78
Lamp black, 113
Lanceolate, 13
Larvae, 123, 153
Larvae third instar, 146
Larvae, first instar, 147
Larval moulting, 122
Larvicide, 146, 147
Laticifers, 18
Latin America, 125
Lead, 143
Leaf disks, 39
Leaf injury index, 143
Leaf mulch, 44
Leaf spotting blight, 46
Leaf web blight, 46
Leather technology, 9
Lemon grass, 147
Lenticular orifice, 61
Lep, 77
Lepidoptera, 124
Leprosy, 82, 83, 98
Leprotic lesions, 82
Leprotic ulcers, 80
Lesions, 95
Leucine, 135
Leucoderma, 82
Leukemia, 95
Leukocytes, 93
Leukocytosis, 116
Leukorrhea, 99
Libriform, 61
Lice, 4, 82
Lignin, 27
Limba, 11
Limbo, 11
Limonoids, 23, 27, 90, 92–96, 117, 123, 159, 160
Linoleic acid, 59, 134
Liperus caponis, 111
Lipids, 19, 130, 141
Lipophilic, 97
Liquorice, 80, 113
Liver function, 116
Loboschiza koenigiana, 46
Lok parampara, 79
Long pepper, 87

Loss of memory, 84
Luminal epithelium, 116
Lyme disease, 146
Lymnaea acuminata, 148
Lymphocytie, 95, 97
Lysimeter, 44
Lysine, 27, 74

Macrophages, 98
Madhuca latifolia, 143
Magnaporthe salvinii, 129
Maha tikatam ghritam, 78
Maha tikatam kashyam, 78
Mahogany, 62
Mahogany, Indian, 62
Majun juzam, 83
Malaria, 9, 82, 115
Malarial parasite, 94
Malaysia, 1, 40, 41
Malham, 77
Mali, 41, 45
Malonic acid, 27
Mannose, 27
Marathi, 11
Margoculture, 33
Margosa, 1, 11, 83
Margosa indica, 1
Margosa oil, 1
Margosan-O, 2
Margousier, 11
Marham bawasir Jadid, 83
Mauritius, 41
Microsporum canis, 92
Mealy bug, 46
Measle, 96
Medullary rays, 58, 61, 83
Melania scabra, 148
Melia azadirachta, 9, 10, 83, 165
Melia azedarach, 2, 10, 18
Melia indica, 9
Meliacins, 130
Membrane permeability, 93, 117
Mennopon gallinae, 111
Mental confusion, 115
Mesocarp, 16, 56

Metabolic acidoses, 116
Metabolizable energy, 133, 135
Metamorphosis, 122
Metatracheal, 61
Methionine, 74, 135
Methylated spirit, 152
Methylenecycleartenol, 27
Metritis, 84
Mevalonate, 38
Mexico, 41
MIC (Methyl iso-cyanate), 141
Micro-cuttings, 39
Microbial spoilage, 72
Micromulsion, 166
Microorganisms, 123
Micropyle, 15
Microsporangial, 39
Microtermes obesi, 148
Middle east, 11
Miscarriage, 84
Miscella refining, 70
Mitochondrial functions, 117
Mitotic, 94
Mixed function oxidases, 116
Mohandjodaro, 1
Mollusc, 122
Molluscicidal, 148
Monosiphonous, 15
Moringa oleifera, 143
Mosquito repellent, 113, 160, 147
Mosquito repellent mat, 147
Moulting, 123, 124
Moulting cycle, 124
Mozambique, 41
Mucus membrane, 81, 137
Multiple shoot, 39
Mung bean, 152
Muriate of potassium, 130
Musca autumnalis, 146
Musca domestica, 146
Muscle glycogen, 88
Mustard oil, 113
Mutagenicity, 134
Myanmar, 9, 11, 41, 74
Mycelia sterelia, 45

Mycoflora, 154
Mycorrhizal, 11
Myristic acid, 59, 134

NAA (Naphthalene acetic acid), 37, 38, 39
NADH (Nicotinamide adenine dinucleotide phosphate), 94
Nasayam, 77
Necrobiosis, 61
Nectar, 80
Neem bitters, 163
Neem coated urea, 129, 130
Neem decline, 45
Neem disorder, 45
Neem gauch, 11
Neem mission, 72
Neem terpenes, 18
Negative chronotropic effect, 96
Nematodes, 3, 82, 122–125, 129, 159–160
Nematodocidal, 87
Nerium indicum, 80
Neta, 77
Neuroendocrine, 122
Neurosecretory proteins, 124
Neurotransmitters, 97
Nicaragua, 25, 41
Nicotine, 95
Nigeria, 25, 41, 42, 151, 152
Nigerian, 11
Nimanuv, 11
Nimba, 77
Nimba arisht, 82
Nimba asav, 82
Nimba ghrit, 82
Nimba haridra, 82
Nimbadi anjan, 82
Nimbadi jatailam palit, 78
Nimbadi kashyam, 78, 92
Nimbadi kwath, 82
Nimbamu, 11
Nimbarishta, 98
Nimbatiktam, 95
Nimbatwagadi kashyam, 97
Nimbay, 11

Nimbdin, 95
Nimbidin, 24, 88, 95–98, 115
Nimbidinic acid, 115
Nimbin, 23, 24, 38, 92
Nimbinin, 92
Nimbiodol, 92
Nimbolide, 38
Nimboloid, 93
Nimbu, 11
Nitrification, 129
Nitrification inhibitor, 65, 130
Nitrifying bacteria, 129
Nitrobacter, 129
Nitrogen peroxide, 141
Nitrogenous fertilizers, 3, 129
Nitrosomonas, 129
Nodular structure, 39
Nonhormonal, 90
Nonopioid, 93
Normoglycemic, 88
Nucellus, 39
Nymph, 123

Obesity, 82
Ocimum sanctum, 113
Octyl gallate, 72
Odontotermes spp., 148
Odoriferous principles, 134
Oedema, 92
Oestrogenic, 90
Oidium azadirachtae, 46
Oil bordeaux, 111
Oil globule, 61
Oil resins, 151
Oleic acid, 59, 134
Olein, 74, 160
Oleodipalmein, 60
Oleopalmitostearin, 60
Opioid, 93
Oral hygiene, 81, 113
Organic acids, 129
Organic matter, 133
Organogenesis, 38, 39
Oriya, 11
Oryzaephilus surinamensis, 45

SUBJECT INDEX

Ovaricotomosed, 90
Ovicidal, 90
Oviposition, 148, 151, 153
Oviposition deterrent action, 124
Ovulation, 89

Palisade cells, 142, 143
Palit Nasyam, 78
Palmitic acid, 59, 134
Palmitodiolein, 60
Panch nimba churnam, 78
Panch tiktam ghritam, 78
Panchamrit, 82
Panchang, 1, 77
Panchnimba avleh, 82
Panchnimba churn, 82
Panchnimba gutika, 82
Panchtikata ghrit, 82
Panchtikata ghrit guggal, 82
Pancreatectomized, 88
Pancreatic beta cells, 88
Papain, 113
Paracetamol, 97
Parasitemia, 93
Paratracheal, 61
Paratracheal zonate, 61
Parietal placentation, 15
Parlatoria orientalis, 46
Particle board, 159
Pathogenic fungi, 123
Patoladi kvath, 78
Peanut cake, 135
Penicillium, 45
Pencillium expansum, 154
Pencillium italicum, 154
Pennisetum purpureum, 136
Peptide glycans, 97
Persian, 11
Pethidine hydrochloride, 92
Phagocytic activity, 98
Phaladi kvath, 78
Pharmacopoeia of India, 78
Pharmacopoeia, Indian, 2
Pharmacopoeial list, 87
Phlegm, 81

Phellogen, 58, 59, 83
Phenol formaldehyde, 159
Phenolic compounds, 79
Phenylalanine, 135
Pheromones, 123
Philippines, 41
Phlebitomus argentipes, 148
Phlebitomus paptsi, 148
Phloem, 57
Phloem parenchyma, 58
Phoma joylana, 46
Phorate granules, 130
Phyllanthus emblica, 78
Phyllotaxy, 13
Phytopesticide, 121
Phytophthora cinnamoni, 12, 45
Phytotoxicity, 142, 143
Pichumarda, 77
Pig, 136
Piles, 79, 83
Pimples, 79, 83
Pin worm, 82, 87
Pine needles, 44
Piperonyl butoxide, 71
Piper longum, 87
Placentation, 15
Plague, 80
Plasmodium berghei, 93
Plasmodium falciparum, 93, 94
Plasmodium yoelli nigeriensis, 94
Plastoglobuli, 142
Plutella xylostella, 153
Pollution, 141
Pollution prevention, 4
Polyethylene glycols, 160
Polymorphonuclear leukocyte, 93, 97
Polyporus, 45
Polysaccharides, 92, 95, 141
Pongamia glabra, 162
Pongamia pinnata, 82, 153
Portuguese, 11, 79
Postparturition disorders, 80
Potassium nitrate, 81
Potassium phosphate, 154
Poultry, 137

SUBJECT INDEX

Praneem, 96
Premature greying of hair, 82
Preservation of food, 4
Preservatives, 62
Progesterone, 90
Progestinol, 90
Prolate spheroidal, 15
Proline, 135
Proline metabolism, 142
Propyl alcohol, 69
Propyl gallate benzophenone, 72
Propyl paraben, 72
Propylene glycol, 98
Prosopis cineraria, 44, 63
Prosopsis juliflora, 42
Prostaglandin, 94
Prostephanous truncatus, 153
Protease, 92, 99
Proteus spp., 88
Protozoa, 93, 94, 111
Protozoal diseases, 94
Pruritus, 111
Pseudococcus gilbertensis, 46
Pseudomonas solanacaerum, 45
Psoriasis, 82, 98
Pubescent, 15
Punaravadi kshyam, 78
Punravadi kvath, 78
Purging, 115
Purification of blood, 83
Pustules, 80, 82
Pyrethroids, 147
Pyrogen-induced hyperpyrexia, 92
Pyrolytic degradation, 73
Pyronimin, 73
Pythium spp., 45

Qatar, 41
Quassia, 80
Queensland, 41
Quercetin, 92, 94
Quick lime, 111
Quinine, 79
Quinine hydrochloride, 91
Quinine sulphate, 93

Rabbit, 137
Rat, 137
Ravisambha, 77
Recurrent generalised seizures, 116
Recurring miscarriage, 84
Repellent, 145, 146
Resin gland, 18
Resin secretions, 18
Resin-secreting glands, 12
Respiratory problems, 82
Retuculitermes speratus, 148
Reye's syndrome, 116
Rheumatism, 82, 84, 92
Rheumatoid disorders, 78
Rheumatoid arthritis, 78
Rhizoctonia solanii, 45, 46, 129
Rhizosphere, 11
Rhodnius prolixus, 94
Rhyzopertha dominica, 152
Ringworm, 82, 98
RNA, 95
Roghan neem, 83
Roller mill, 66
Rooting, 37
Round worm, 87
Rubber industries, 160
Russel viper, 80
Rutales, 123

Sahara, 41
Sahel, 42
Salanin, 26
Salannin, 130, 165
Salmonella paratyphi, 88
Salmonella typhi, 88, 100
Salmonella typhimurium, 117
Samvardhan samithi, 79
Sanskrit, 77
Sapindus mukrosii, 91
Sarcoma, 95
Sarcoptes scabei, 99, 111
Sarcoptes scabei var. *caprae*, 111
Sarcoptic mange, 111
Saturated glycerides, 59
Saudi Arabia, 41

SUBJECT INDEX

Scabies, 98, 99
Scanning Electron Microscope, 61
Schistocera gregaria, 24
Schizo-lysigenous cavities, 19
Schizonticidal, 93
Sclerenchyma, 83
Sclerotium oryzae, 129
Scrofula, 98
Scytalidium anamorph, 99
Secondary bast, 83
Secondary metabolites, 79
Secretory cavities, 57
Secretory cells, 18
Secretory system, 124
Sedative, 96
Seed development, 16
Seedling morphology, 16
Senegal, 41
Senile dementia, 84
Septicemia, 100
Serine, 135
Serum acid phosphate, 89
Serum cholesterol, 116
Serum value, 116
SGOT, 116, 136
SGPT, 136
Shay ulceration, 95
Sheep, 137
Sheetal, 77
Shelter bed, 42
Shivalik, 9
Shoot bud, 39
Shuttle, 62
Sialogogue, 87
Siddha, 1
Sidha system, 98
Silicates, 160
Silviculture, 42
Silvopastoral, 46
Silvipasture, 46
Sinhali, 11
Sitophilus oryzae, 151, 153
Sitophilus zeamais, 152
Sitotroga cerealetta, 152
Skin diseases, 80, 83

Skin encrustation, 111
Sleeplessness, 84
Snail, 148
Snake bites, 80
Snake-skin, 9
Soap nut, 91
Sodium aluminium sulfate, 79
Sodium and potassium margosate, 98
Sodium chloride, 42, 73
Sodium hydroxide, 131, 160
Sodium nimbidinate, 89, 92, 96, 115
Sodium nimbinate, 89
Soft wood cuttings, 38
Soil erosion, 43
Solanum indicum, 78
Solenopsis, 45
Solerotium rofsii, 45
Sore throats, 80
South Pacific countries, 125
Spathulate, 15
Spawning period, 81
Specific gravity, 62
Sperm, 89
Spermatogenesis, 89–91
Spermicidal, 89–91, 112
Spermicide, 165
Spinach, 81
Spleen cells, 99
Spongy gums, 79
Spongy tissue, 18
Sporostatic, 91
Sri Lanka, 1, 11, 41, 151
Staminal filament, 15
Staphylococcus aureus, 88
Staphylococcus citrus, 88
Staphylococcus epidermidis, 88
Staphylococcus lactitis, 88
Stearic acid, 59, 134
Stearin, 74
Stearodiolein, 60
Stereochemistry, 25
Steroids, 23, 90
Sterols, 94
Steror, 115
Stimulant, 87

Stimulative cerebral tonic, 82
Stomata, 58
Stomatal conductance, 142
Stomatal frequency, 141
Stomatitis, 137
Stool, hard, 84
Stratum corneum, 115
Streptozotocin, 89
Structural degradation, 24
Strychnos nux vomica, 98
Stupor, 115, 116
Styrene, 160
Sub-prolate, 15
Succinic acid, 27
Sudan, 24, 41, 42
Sulfur dioxide, 141, 142
Sulphonylureas, 88
Superoxide dismutase, 142
Suppositories, 91
Surfactants, 160
Surinam, 41
Swellings, 78
Symptomatology, 84
Syncarpous, 15
Synergids, 15
Synthetic pesticides, 121
Syphilis, 79, 98

Tachypnoea, 116
Thai neem, 17
Tamabin, 11
Tamil, 11
Tannery, 141
Tanning effluents, 73
Tannins, 79, 130
Tanzania, 41
Tar, 146
Tea mosquito, 46
Teak, 62
Tectona grandis, 62
Teli, 67
Telugu, 11
Tendency to miscarriage, 84
Tensile strength, 62
Terminalia belerica, 78

Terminalia chebula, 78
Termite, 148
Terpenic compound, 95
Terpenoid, 93
Testes, 91, 94, 137
Testicular function, 91
Testosterone, 91
Tetanus, 98
Tetraplema tetrapera, 72
Tetrasulphides, 27
TH-1 component, 98
Thailand, 1, 17, 40, 41
Theileria annulata, 111
Thermal electric plants, 159
Thermal insulation, 159
Thiabendazole, 154
Thiktakam ghritam, 78
Thiktakam kashyam, 78
Thinbaw, 11
Thioamyl alcohol, 27
Third instar larvae, 146
Threadworm, 87
Threonine, 27, 135
Throat problems, 82
Ticks, 111
Tilapia fish, 153
Tinea pedis, 99
Tinospora cordifolia, 78
Tissue culture, 38
Tissue-repairing, 111
Tobacco, 160
Togo, 41
Tonic, 82
Toxicity, 11
Toxicology, 115
Tracheal mites, 143
Tracheids, 61
Transplant shock, 11
Treachery elements, 61
Tremors, 133
Tremulous gait, 116
Tribolium castaneum, 152, 153
Tribolium confusum, 152
Trichophyton mentagrophytes, 92, 99
Trichosanthes dioica, 78

Trilacunar, 16, 17
Trilocular, 15
Triple fusion, 15
Triterpenoid, 18, 90, 99
Triunsaturated glycerides, 59
Trogoderma granarium, 152
Trigonella foenum graceum, 34
Trypanocidal, 94
Trypanosoma bruci, 94
Trypanosoma cruzi, 94
Trypanosoma gambiensis, 94
Tsetse, 94
Tulsi, 113
Tumor cells, 38
Tumors, 38
Turmeric, 82, 153
Twig canker, 46
Twin roller, 71
Typhoid fever, 84
Tyrosine, 135

Uganda, 41
Ulcerogenic, 95
Ulcers, 78, 80, 83
Ultra violet, 122
Ultra violet radiation, 24
Unani tibb, 1, 82, 87
Unilacunar, 16
United States, 154, 163
Unsaturated hydroxy aldehyde, 117
Urea neem cake, 130
Urease enzyme, 130
Uric acid, 116
Urinary disorders, 82
Urinary passage, 83
Urtica, 98
US National Formulary, 2
User, 42
Uterine muscle contraction, 80
Uterus, 90
Uttar Pradesh, 9

Vaccinia virus, 96
Vagina, 81, 90, 99
Vaginal mucosa, 91

Vaginal smear, 90
Vaginal tissue, 90
Vaginitis, 99
Vaipilla, 81
Vaipilla tailam, 81
Valine, 135
VAM (Vesicular arbuscular mycorrhizal infection), 11
Vegetative propagation, 37
Varanejatayidi ghritam, 78
Variola virus, 95
Vascular permeability, 92
Vasodilation, 96
Vatika, 77
Vector, 146
Venereal disease, 79, 83
Venom, 80
Vepa, 11
Vepenin, 23
Veppam, 11
Verticillum, 45
Vesicles, 18
Vessels, 61
Vietnam, 41, 152
Vigna radiata, 153
Vigna unguiculata, 152
Vilasinin, 23
Village pharmacy, 1
Vimlu, 11
Viral hepatitis, 96
Virus, 125, 147
Vitex negundo, 153
Vitiligo, 99

Wasteland, 42, 43
Weeping eczema, 98
Weevils, 151
West Indies, 41
White patches, 82
Wind breaker, 3, 43, 44
Winnower, 67
Winnowing, 71
Woodfordia fruitcosa, 98
Worm infestation, 82
Wound sepsis, 100

Xanthmonas azadirachti, 46
Xylaria azadirachtae, 45
Xylose, 27

Yaman, 41
Yeppa, 11
Yograja guggulu, 78

Zimad bawasir, 83
Zimad mohasa, 83
Zinc, 135
Zingiber officinale, 78
Zygote, 15, 90